BASIC GEOLOGICAL MAPPING

The Geological Society of London Handbook Series
published in association with
The Open University Press comprises

Barnes : *Basic Geological Mapping*
Tucker : *The Field Description of Sedimentary Rocks*
Fry : *The Field Description of Metamorphic Rocks* (in preparation)
Thorpe : *The Field Description of Igneous Rocks* (in preparation)

Basic Geological Mapping

John W. Barnes

Department of Geology
University College of Swansea

THE OPEN UNIVERSITY PRESS
MILTON KEYNES
and
HALSTED PRESS
John Wiley & Sons
New York – Toronto

First published 1981 by
The Open University Press
A Division of
Open University Educational Enterprises Limited
12 Cofferidge Close
Stony Stratford
Milton Keynes, MK11 1BY
England

**Printed in Great Britain by
Redwood Burn Ltd, Trowbridge**

British Library Cataloguing in Publication Data:

Barnes, John W.
Basic geological mapping. – (Geological Society of London handbook series; 1)
1. Geological, mapping
I. Title II. Series
550 Q36

ISBN 0-395-10035-X (Open University Press)

Published in the U.S.A., Canada and Latin America by Halsted Press, a Division of John Wiley & Sons, Inc., New York.

Library of Congress Cataloguing in Publication Data :

Barnes, J.W. (John Wykeham), 1920–
Basic geological mapping.

(Geological Society of London handbook series)
Includes bibliographical references.
1. Geological mapping. I. Title. II. Series.
QE36.B33 1982 526 81-7151
ISBN 0-470-27250-3 **(Halsted)** AACR2

Contents

ERRATUM

Figure 5.13 on page 63 should be inverted.

Preface

This book is a *basic* guide to field techniques. It is meant to be kept in camp with you and even, at times, to be carried in your rucksack in the field. In addition, because no piece of geological mapping can be considered complete until the geology has been interpreted and explained, chapters are provided on drawing cross-sections; on preparing and presenting 'fair copy' maps; and on preparing geological diagrams from your fieldwork suitable for inclusion in your report. Report writing itself cannot be covered here, but it must be borne in mind that a report explaining the geology is an essential part of any field project. Some emphasis, too, is given to field sketch mapping because many reports lack those large-scale detailed maps of small areas which can often explain complex aspects of the geology that cannot be shown on the normal scale of map being used, and which are difficult to describe in words. Attention is also given to field notebooks which are, in many cases, deplorable.

It is assumed that readers of this book have already had at least one year of university or equivalent geology and have been told what to look for in the field. Geological mapping cannot, however, be taught in lectures and laboratory: it must be learnt in the field. Unfortunately, only too often, trainee geologists are left largely to their own devices, to sink or swim, and to learn to map for themselves with a minimum of supervision on 'independent mapping projects'. It is hoped that this book, together with others in the same series, will help them in that task.

John W. Barnes
1981

Acknowledgements

It is impossible to name everyone who has been associated with the production of this book. Among those who deserve special mention for help with text or diagrams are D.V. Ager, F.R. Cross, A.R. Gardiner, R.H. Graham, T.R. Owen, S.J. Matthews and P. Styles, all of the University College of Swansea; M. H. de Freitas of Imperial College; and also my colleague on ten international mapping projects, E. H. Bailey of U.S.G.S. Appreciation is also due to departmental drawing office staff and particularly to Mrs V.M. Jenkins who typed the drafts.

1 *Introduction*

There are many kinds of geological map, from small-scale reconnaissance surveys to detailed underground mine maps and large-scale engineering site plans, and each needs a different technique. This book, however, is concerned only with the rudiments of geological mapping. Its intention is to give the young geologist a basic knowledge upon which he can build. It cannot tell him everything he needs to know but it is hoped that it will stimulate his imagination so that he can adapt his methods to prevailing field conditions and develop and invent new methods when necessary. A geologist must also remember that *accurate* geological maps are the basis of *all* geological work, even laboratory work, for it is pointless to make a detailed investigation of a specimen whose provenance is uncertain. As Wallace (1975) said in his 1974 Jacklin lecture: 'There is no substitute for the geological map and section—absolutely none. There never was and there never will be. The basic geology still must come first—and if it is wrong, everything that follows will probably be wrong.'

1.1 Outline and approach

This book is arranged in what is hoped is a logical order for someone about to go into the field on his first 'independent mapping' project. First, it describes the equipment he will need; then he is introduced to the many types of geological maps he may have to deal with at some time during his professional career. A description of the different kinds of topographic base maps which are available to him and on which he may plot his geology in the field follows. Methods of locating himself on his maps are also described and advice is given on what to do when no maps at all are obtainable.

The next three chapters describe the methods and techniques of geological mapping, including a brief description of photogeology. Another chapter is devoted to the use of field maps and those much neglected items, field notebooks.

The last two chapters concern 'office work', some of which may be done in your field camp. They cover methods of drawing cross-sections and the preparation of other diagrams to help interpretation; including the construction of 'egg-crate' models which make complex structures easier to visualize. Advice is also given for preparing a 'fair drawn' copy of your field map, and on illustrating your report on the geology mapped, for no mapping is complete until the geology has been explained. A geological map is not, as sometimes supposed, an end in itself. The object is to *explain* the geology and the map is only part of that process.

The approach throughout is practical. This is a 'how to do it' book and avoids any theoretical considerations of geology. Those can be found elsewhere. The object is to tell the reader what to do in the field to collect the evidence from which to draw his conclusions. What conclusions he draws is up to the reader himself.

1.2 Field behaviour

Geologists spend much of their time in the open air and more often than not their work takes them to the less inhabited parts of a country. If they did not like open country one presumes they would not have become geologists, consequently, it is taken for granted that geologists are conservation minded and have a sympathetic regard for the countryside and those who live in it: do not leave gates open, climb dry stone walls or trample crops, and do not leave litter or disturb communities of plants and animals. When you are collecting specimens, do not strip or spoil sites where type fossils or rare minerals occur. Take only what you need. Always ask permission to enter land from the owners, agents or authorities unless it is specifically known to be open to the public. Most owners are willing to cooperate if they are asked to but are understandably annoyed to find strangers sampling their rocks uninvited. Geologists should bear in mind that upset landowners can inhibit geological activities in an area for years to come, and this has already happened in parts of Britain. Many other countries are less overpopulated and have more open space, and the situation may be easier, but every country has some land where owners expect people to consult them before working there. If in doubt, ask!

1.3 Safety

A geologist must be fit if he is to do a full day's work in the field, perhaps in poor weather or a difficult climate. Geological fieldwork, in common with other outdoor pursuits, is not without physical hazards. However, most risks can be minimized by following fairly simple rules of behaviour, and discretion may often be the better part of valour, say, when faced with a rock exposure in a difficult position. A geologist is very often on his own, with no one to help him should he get into difficulties. Experience is the best teacher but common sense is a good substitute. Field safety is discussed further in Appendix I from the standpoint of both the student (or employee) and his supervisor (or employer).

Finally, a few words of comfort for those about to start their first piece of independent mapping. The first week or so of nearly every geological mapping project can be depressing, especially when you are on your own in a remote area. No matter how many hours are spent in the field each day, little seems to show on the map except unconnected fragments of information which have no semblence to an embryo geological map. Do not lose heart: this is quite normal and the map will suddenly begin to take shape. The last few days of fieldwork are also often frustrating for, no matter what you do, there always seems to be something left to be filled in. When this happens, check that you do have all the essential information and then work to a specific finishing date. Otherwise you never will finish your map.

2

Instruments and equipment

A geologist needs relatively little equipment in the field. A hammer is essential, so is a compass, clinometer, short steel tape and a handlens. He also needs a map case, notebook, map scale, protractor, pencils and eraser, plus an acid bottle and a pen-knife. On occasion, he needs a 30 m tape, an altimeter, a pedometer, a stereonet and a pocket stereoscope. Felt-tipped pens and timber crayons are excellent for marking specimens; chisels and moils may be needed to break them off. Although not essential, a camera and binoculars can be helpful. Finally he needs a rucksack to carry everything in, including, of course, his lunch.

A geologist must also wear proper footwear and clothes if he is to work efficiently, often in wet, cold weather, when other people stay indoors. Poor clothing may even put him at risk. Make a checklist of what you need to pack for field trips and always refer to it before setting out for the field.

2.1 Hammers and chisels

Any geologist going into the field needs at least one hammer with which to break rock. Taken generally, any hammer weighing less than 1½ lbs (0.7 kg) is of little use except for very soft rocks: 2–2½ lbs (0.9–1.2 kg) is probably the most useful weight. The commonest pattern still used in Europe has one square-faced end, and one chisel-end. Geologists associated with the mining industry tend to favour a 'prospecting pick': it has a long pick-like end whch can be inserted into cracks for levering out loose rock, and can also be used for digging into soil in search of float. Both types can be bought with a choice of wooden or fibreglass handles, or with a steel shaft encased by a rubber grip (Fig. 2.1). If a wooden handle is chosen, buy spare handles and some iron wedges to fasten them on with.

Geologists working in granite and gneissic regions may opt for heavier hammers. Four-pound (1.8 kg) geological hammers are available but a bricklayer's 'club' hammer, shaped like a small sledge hammer, can be bought more cheaply. It is even more effective if its rather short handle is replaced by a longer one.

Hammering alone is not always the best way to collect rock or fossil specimens. Sometimes a cold chisel is needed to break out a specific piece of rock or fossil. Its size depends on the type of work to be done. A ¼-inch (or 5 mm) chisel may be ideal to chip a small fossil free from shale, but for breaking out large pieces of hard rock a ¾- to 1-inch (or 20 mm) chisel is

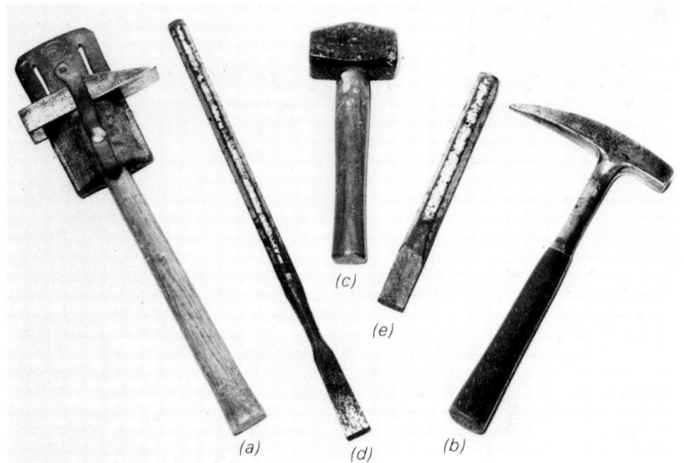

Fig. 2.1 Tools for the field: (a)—traditional geologist's hammer in leather belt 'frog'; (b)—steel-shafted 'prospecting pick'; (c)—bricklayer's 'club' hammer; (d)—45 cm long chisel with 2.5 cm edge; (e)—2.5 cm chisel 25 cm long.

needed (Fig. 2.1). Perhaps geologists should follow the lead of mine samplers who find a 'moil' more effective. This is a steel bar, 25–30 cm long and 2.5 cm diameter, which has been sharpened to a point and tempered. One thing a geologist should never do, however, is to use one hammer as a chisel, and hit it with another. The tempering of a hammer is quite different from that of a chisel: a hammer face is hardened, a chisel top is not, and small steel fragments may fly off the hammer face with unpleasant results.

Some geologists carry their hammers in a 'frog' or hammer holster, as this leaves their hands free for climbing, writing and plotting. Hammer frogs are essential when mapping underground mine workings where ladders must be climbed. They can be bought or easily made from heavy leather (Fig. 2.1). Note that the use of a geological hammer is a 'chipping action', an operation specifically mentioned by the British *Health and Safety at Work Act* as needing the use of approved goggles. Courts would probably take a less than liberal view of claims for compensation for eye injuries suffered when goggles were not being worn.

2.2 Compasses and clinometers

The ideal geologist's compass has yet to be designed. Americans have their *Brunton*, the French the *Chaix Universelle*, the Swiss have the *Meridian*, but many geologists now use Swedish *Silva* compasses (Fig. 2.2). These are reasonably priced, well-damped needle compasses. The 15TD.CL model incorporates a

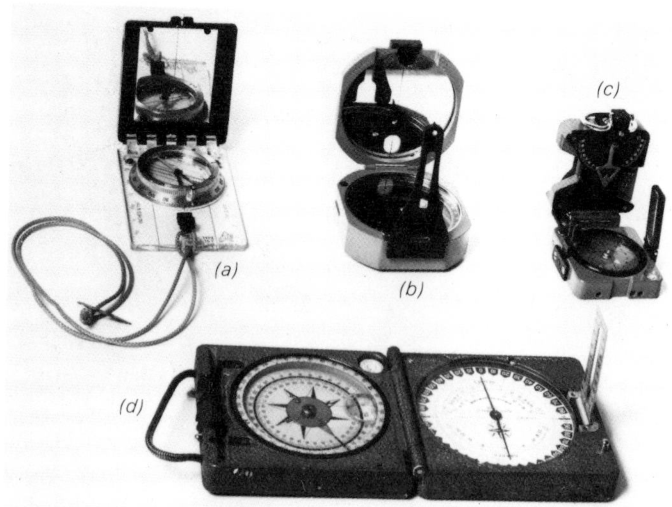

Fig. 2.2 Compasses designed for the geologist: all incorporate a clinometer. (a)—Swedish *Silva* compass type 15TD.CL; (b)—American *Brunton* 'pocket transit'; (c)—Swiss *Meridian* compass; (d)—French *Chaix-Universelle*. The Brunton and Meridian can be used as hand-levels.

clinometer and has been designed specifically for geologists. One convenience is that bearings can be plotted onto the map immediately after they have been measured by using the compass itself as a protractor (see Section 5.2). Like the Brunton, Silva needle compasses are not ideal for sighting distant points and the makers now produce a neat card-type prismatic model (No. 54) but it has no clinometer (Fig. 2.3). The drawback of nearly every compass available, except for the Brunton and the Swiss Meridian is, however, the lack of a hand-levelling device, so necessary for many geological tasks.

2.2.1 Compass graduations

Compasses can be graduated in several ways. The basic choice is between the traditional 360° (degrees) and the continental 400g (grads) to a full circle. Both are used in continental Europe. If you opt for degrees, you must then choose between graduation of the compass into four quadrants of 0°–90° each, or to read a full circle of 0°–360° ('azimuth' graduation). The writer recommends azimuth, for bearings can be expressed more briefly and with less chance of error. Comparisons are made in Table 2.1.

2.2.2 Using compasses

Prismatic compasses and mirror compasses are used in different ways when sighting a distant point. A prismatic is held at eye level and aimed like a rifle so that the point is aligned

Table 2.1

Quadrant bearing	Azimuth bearing
N 36° E	036°
N 36°W	324°
S 36°E	144°
S 36° W	216°

with the compass fore-sight and the slot in the top of its prism; the bearing is read simultaneously through the prism. A mirror compass, such as a Brunton or Silva, is held at waist height and the distant point aligned with the front-sight so that both are reflected in the mirror and bisected by its hair-line (Fig. 2.4). If the compass is not liquid damped, the bearing is read when the needle has *almost* settled by averaging the limits of its swing. A mirror compass can be used, if awkwardly, in a similar manner to a prismatic by holding it at eye level and reflecting the needle in the mirror. Mirror compasses have a specific advantage over prismatics in poor light, such as underground.

2.2.3 *Clinometers*

Not all compasses incorporate a clinometer into their construction. Clinometers can be bought separately and a few types, such as the Finnish *Suunto*, have the advantage that they can also be used as a hand-level. Some hand-levels, such as the *Abney*, can also be used as a clinometer, but rather

Fig. 2.3 Various other compasses: (a)—Japanese *Europleasure Lensatic* compass. This can be read like a prismatic compass, and is liquid filled; (b)—standard British prismatic compass made by a variety of manufacturers. They are robush and liquid filled. (c)—Swiss *Meridian* liquid-filled prismatic compass; (d)—Swedish *Silva* liquid-filled prismatic compass, No. 54; (e)—Japanese '*universal clinometer*' an adaptation of the traditional miner's 'hanging compass': for the measurement of lineations (see also Fig. 2.6).

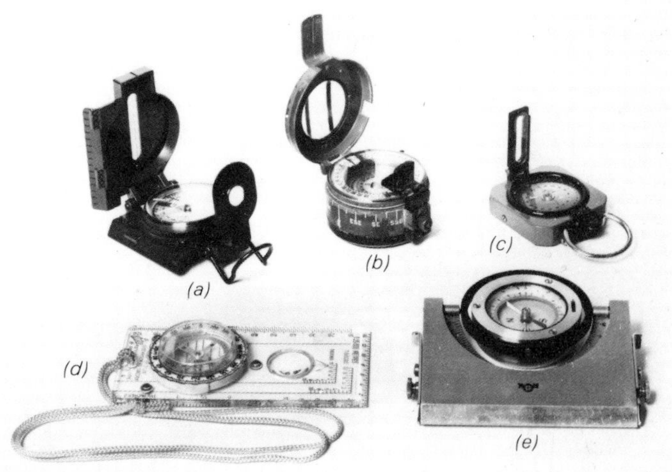

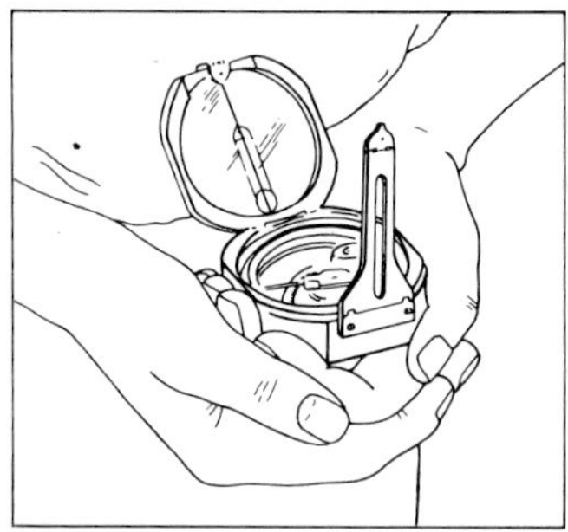

Fig. 2.4 Correct way to hold a Brunton (or any other mirror-type compass) to sight a distant point (reproduced by courtesy of the Brunton Company, Riverton, Wyoming, USA).

inconveniently. The *Burgess* 'level and angle indicator', designed for do-it-yourself handymen, makes a cheap and effective clinometer. A builder's 'two-foot' clinometer rule is useful for measuring lineations. Clinometers can be made, either by using the pendulum principle or, more easily, as follows: stick half a semicircular protractor—zero-point downwards—to a rectangle of perspex and cement a piece of plastic (windscreen-washer) tubing around its edge: Insert a small bead into the tube and plug its ends: the bead acts as a point as it rolls round the tube (devised by G. Bryn Thomas). A selection of clinometers is shown in Figure 2.5.

2.2.4 *Lineation compass*

The Japanese produce a compass designed to measure trend and plunge simultaneously. The compass case is in gimbals so that it always remains level whatever the angle of its framework. It is effective even in the most awkward places (Figs. 2.3e and 2.6).

2.3 Handlenses

Every geologist must have a handlens and should develop the habit of

Fig. 2.5 A selection of clinometers: (a)—*Burgess* 'level and angle indicator'; (b)—home-made clinometer; (c)—builder's 'two-foot' rule with level bubble and hinge graduated at 5° intervals; (d)—*Abney* hand-level—can be used as a clinometer too.

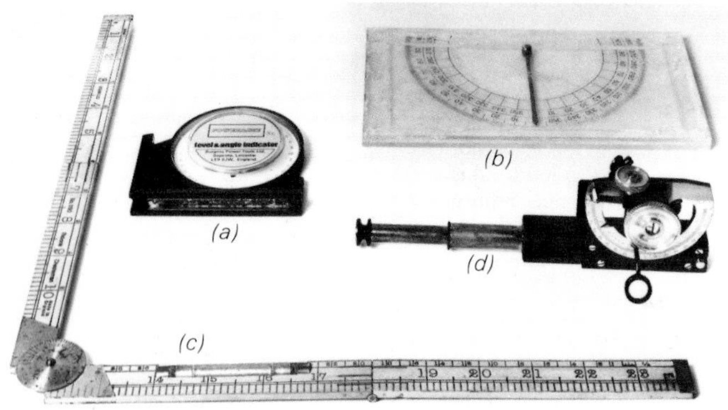

Fig. 2.6 Japanese *universal clinometer.* Trend can be read directly off the compass and plunge from the pointer which can be seen below the compass box. It can be used in many awkward places, including overhangs (made by Nihon Chikagakso Shaco, Kyoto).

carrying it at all times so that when he needs it, he has it with him. A magnification of from 7 to 10 times is probably the most useful. Although there are cheap ones on the market, a good quality lens is worth the extra cost in flatness of field, and should last a lifetime. To ensure that it does last a lifetime, attach a thin cord to it to go round your neck.

2.4 Tapes

Always carry a short 'roll-up' steel tape. A 3 m tape takes up no more room than one 1 m long and is only slightly heavier and much more useful. A geologist occasionally also needs a 10 or 30 m 'linen' tape for small surveys. He might not need it every day but should keep one in camp for when he does. Treat tapes with respect. Wind them back into their cases only when clean, for dirt will wear off their graduations. If a linen tape is muddy, coil it into loops between measurements. When you do eventually wind it back into its case, do so between two fingers of the left hand or through a damp rag, to wipe off the dirt. When finished for the day, wash and dry it before putting it away.

2.5 Map cases

A map case is obviously essential where work may have to be done in rain or in mist; but even in warmer climes, protection from both the sun and sweaty hands is still needed. A map case must have a rigid base so that the map can be written on and bearings plotted; it must protect the map; and it *must* open easily, otherwise it will deter you from adding information to your map. The best are probably home made (Fig. 2.7). Pencil holders make mapping easier whether attached to your map case or your belt. Make your own.

Fig. 2.7 A map case made from a perspex sheet attached to a plywood sheet by a nylon (or brass) 'piano hinge'. The hinge is fixed by 'pop-rivets'. A wide rubber band keeps maps flat.

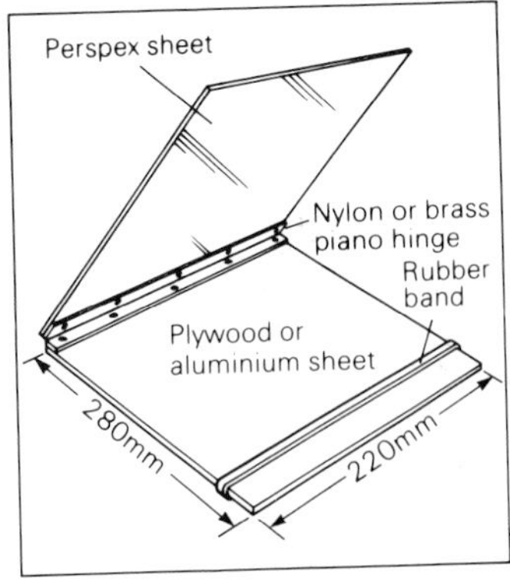

2.6 Field notebooks

Do not economize on your field notebook. It should have good quality 'rainproof' paper, a strong hard cover and good binding. It has to put up with rough usage, often in wet and windy conditions. Nothing is more discouraging than to see pages of field notes torn out of your book by the wind and blown across the countryside. Loose-leaf books are particularly vulnerable. A hard cover is necessary to give a good surface for writing and for sketching. A notebook should fit into your pocket so that it is always available but be big enough to write on in your hand. A goods size is 12 cm x 20 cm. Try to buy one with squared—preferably metric squared—paper; it makes field sketching so much easier. Half-centimetre squares are quite small enough.

2.7 Scales

A geologist must use a proper scale, most conveniently about 15 cm long. A ruler is just not good enough: it seldom has an edge thin enough to allow accurate plotting of distances, and converting in your head a distance measured in metres on the ground to the correct number of millimetres on the scale, leads to errors. Scales are not expensive for the amount of use they get. Most are oval in section and engraved on both sides to give four different graduations. The most convenient combination is probably: 1:50,000, 1:25,000, 1:12,500 and 1:10,000. In Britain a scale with the old 1:10,560 is useful, for 6 inches = 1 mile sheets still exist in many areas. In the USA scales with 1:62,500 and 1:24,000 graduations are needed. Colour code scale edges by painting each with a transparent waterproof ink, or attach coloured adhesive tape, so that the edge you need is immediately recognizable.

2.8 Protractors

Little need be said about protractors. They are easily obtainable and relatively cheap. They should be 15–20 cm in diameter and semi-circular: circular protractors are unsuitable for plotting bearings in the field. Always carry at least two spare smaller protractors (10 cm diameter) to guard against loss. Transparent protractors are difficult to see when dropped but are easier to find if marked with an orange fluorescent adhesive 'spot'. Fig. 2.8 shows a selection of scales and protractors.

2.9 Pencils and erasers

At least three lead pencils are needed in the normal course of mapping: a hard pencil (4H or 6H) for plotting bearings; a softer pencil (2H or 4H) for plotting strikes and writing notes on the map; and a further pencil (2H) for writing in your notebook. The harder alternatives are for warmer climates, the softer for cold. Do not be tempted into using soft pencils; they smudge, and they need frequent sharpening. A soft pencil is quite incapable of making the fineness of line required on a geological map with sufficient permanency to last a full day's mapping in rigorous conditions. Buy only good quality pencils, and when possible, buy them with an

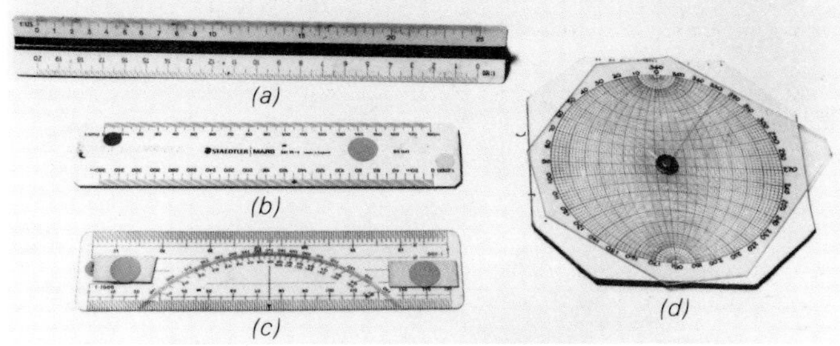

Fig. 2.8 A selection of scales, and protractors: (a)—a triangular map scale which is not recommended for field use; (b)—a plastic scale with graduations on both sides; (c)—transparent combined protractor and map scale (C-thru Ruler Company, Bloomfield, USA); (d)—a home-made pocket stereonet for field use. The upper rotating surface is slightly matt so that it can be drawn on in pencil and easily cleaned. Note that orange fluorescent adhesive 'spots' have been stuck to the scales to make them easier to find if dropped.

eraser attached. Alternatively buy erasers which fit over the end of a pencil. Attach a larger, good quality eraser to your buttonhole or map case by a piece of string, and always carry a spare. Coloured pencils should also be of top quality: keep a list of the make and shade numbers you do use so you can replace them with exactly the same colours again.

2.10 Acid bottles

Always carry an acid bottle in your rucksack. It should contain a *small* quantity of 10% hydrochloric acid. Five ml is usually ample for a full day's work, even in limestone country, providing only a *drop* is used at a time—and one drop should be enough. Those tiny plastic dropping bottles in which some proprietary ear-drops and eye-drops are supplied make excellent acid bottles. They have the advantages that they deliver only one drop at a time, are small, do not leak and will not break.

2.11 Other instruments

These are listed below in the order they are most likely to be used.

2.11.1 Stereonets

A pocket stereonet is most useful when mapping lineations. Plunge and trend can be calculated on the spot from strike and pitch measurements made on bedding and foliation planes or from the intersections of planes. A stereonet is the geologist's slide rule and the structural geologist will find many uses for it in the field. Make your own but remember a Wulff equal-angular net not the Schmidt equal area net (Fig. 2.8) is needed for calculating angles and directions.

2.11.2 Stereoscopes

You need a pocket stereoscope when mapping on aerial photographs. It allows you to obtain a 3-dimensional stereoscopic image from stereo-pairs of photographs in the field.

2.11.3 *Pedometers*

A pedometer is useful only for reconnaissance mapping at scales of 1:100,000 or smaller. It does not actually measure distances directly; it counts paces and expresses them in terms of distance after it has been set to your own pace length. Make allowances for your shorter steps on slopes.

2.11.4 *Aneroid barometers*

There are occasions when an altimeter, i.e. a barometer graduated in altitudes, can be a useful aid. Excellent, robust, pocket-watch sized instruments, such as the Thommen mountain altimeter are sufficiently accurate for many geological purposes and are not particularly expensive. Remember to carry them in your hand baggage when travelling by air: the cabin is pressurized, the baggage compartment is not. As most instruments read only to about 5000 m above sea level, they are unlikely to function properly if exposed to the 10,000 or 15,000 m of modern air travel.

2.12 Field clothing

To work efficiently, a geologist must be properly clothed. In temperate and colder climates, wear loose-fitting trousers: tight jeans are not as warm. In very cold weather, wear 'long johns' or pyjama trousers underneath your field trousers, or put on your waterproof overtrousers. Carry a sweater and a waterproof anorak or cagoule in your rucksack in warm weather, but as it gets colder wear a padded field jacket, a thicker shirt, and a warm vest beneath. Make sure that you always carry waterproof and windproof overclothing for use in bad weather. Take also a woolly hat: heat is lost rapidly through your scalp and it is not always convenient to wear the hood of your anorak. Use gloves in cold weather. Some geologists prefer fingerless mittens so that they can write on their map with them on, but keep a pair of water-resistant gloves with you too. Gloves are probably lost more frequently than any other piece of equipment, so keep a spare pair in camp. Clothing in warm climates is less important, but long-sleeved shirts and long trousers should be worn in the field until you are fully acclimatized to strong sunlight. Sunburn is painful at best, and can be dangerous. A 'jungle hat' is also recommended. The brim pulled down over your eyes is probably a far more effective shade from glare than sun glasses.

Boots for fieldwork in temperate, wet and cold climates should be strong and waterproof, with well-cleated soles. Leather boots are expensive but they are a part of a geologist's essential equipment. Rubber 'Wellington' boots can be worn when working in boggy ground and some types now have excellent soles which allow them to be used on rock. Rubber boots are not, however, comfortable for walking long distances. In warm, dry climates, lightweight half-boots of the 'chukka boot' type, or even hockey boots, are ideal. Heavier boots, however, are still advisable in mountains wherever you are.

When buying field clothing, choose a yellow, orange or red anorak for preference as this is more easily seen by search parties in an emergency.

3

Geological maps and base maps

To make a geological map a geologist requires a topographic base map on which to plot his geology in the field. He also needs a topographic base on which to plot his interpretation of the geology to form the 'fair copy' map he submits to his employer when his work is complete. In Britain, a geologist has 'Ordnance Survey' (OS) maps at his disposal at a scale of 1:10,000, and even larger in many areas. Elsewhere, the scales of the maps available to him are probably smaller; he may even have difficulty in getting a base map at all, for many countries restrict the issue of all but tourist maps to officials. He may even have to make his own topographic map—if he knows how. Any geologist, especially if he intends to enter the mineral industry, is well advised to learn at least the rudiments of map making. It will stand him in very good stead.

3.1 Types of geological map

Geological maps fall into four main groups. These are reconnaissance maps, maps of regional geology, large-scale maps of limited areas and maps made for specialist purposes. Small-scale maps covering very large regions are usually compiled from information selected from one or more of these groups.

3.1.1 Geological reconnaissance maps

A reconnaissance map is made to find out as much as possible about the geology of an unknown area as quickly as possible. It is usually made at a scale of 1:250,000 or less, sometimes very much less. Some reconnaissance maps are made by *photogeology*, that is, by interpreting geology from aerial photographs with only a minimum amount of work done on the ground to identify rock types. Reconnaissance mapping may even occasionally be done by plotting the main geological features onto a base map from a light aircraft or helicopter with, again, only brief confirmatory visits to the ground itself. Airborne methods are particularly useful in regions where field seasons are short, such as in northern Canada and Alaska.

3.1.2 Regional geological maps

Reconnaissance has given the outline of rock distribution and general structure, now the geology must be studied in more detail, most commonly at a scale of 1:25,000 or 1:50,000.

Regional maps should be plotted on a reliable base. Unfortunately, in some countries, geological mapping outstrips topographic coverage and the geologist must survey the topography himself. An accurate geological map loses much of its point if superimposed on an inadequate topographic base.

Regional geological mapping done on the ground may be supported by systematic photogeology and it should be emphasized that photogeological information is *not* inferior to that obtained on the ground although it may differ. Some geological features seen on aerial photographs cannot even be detected on the ground while others can be more conveniently followed on photographs than in surface exposures. Regional geological mapping should incorporate any techniques which can help in plotting geology and which the budget will allow, including geophysics, pitting, augering and drilling.

3.1.3 Detailed geological maps

Here, scale is anything from 1:10,000 upwards and they are usually made to investigate a specific geological problem, perhaps resulting from discoveries made during regional mapping, or perhaps with an economic objective, such as a dam site or mineral investigation.

3.1.4 Specialized maps

Specialized maps are many and varied. They include large-scale maps made in great detail of small areas to record specific geological features. Many are made for economic purposes, such as open pit mine plans at scales from 1:1000 to 1:2500; underground geological plans at 1:500 and even larger; and engineering site investigations at similar scales. There are many other types of maps with geological affiliations too. They include geophysical and geochemical maps, foliation and joint maps, sampling plans with outline geology, maps of drift coverage and maps of the sub-surface. Many are prepared as transparent overlays to be superimposed on a normal geological map at the same scale.

3.2 Topographic base maps

3.2.1 Great Britain

The Ordnance Survey (OS) produces several large-scale topographic map series for geologists to choose from. Particularly useful are those at 1:10,000 (now rapidly replacing the older 6 inches = 1 mile maps) and 1:25,000. The wealth of detail printed on these maps, including hedges and fences, makes the accurate location of geological observations relatively simple except, perhaps, in moorland, where it becomes only a little more difficult.

Maps at still larger scales are available. These are the 1:2500 series, referred to as '25-inch maps' although their true scale is 25.344 inches to a mile, and the 1:1250 series ('50-inch maps'). Neither are contoured but 'spot heights' are shown. The 1:2500 series covers the greater part of Britain except for mountains and moors, but the 1:1250 plans are available only for urban areas with populations over 20,000. Both can be obtained as printed plans on paper; as photographic 'printouts' from *SIM* (Survey Information Microfilm); and as 'copycards' (35 mm film in cardboard mounts) from which you can make your own photographic copies. OS maps can be bought from approved distributors throughout the country who carry maps covering their own areas. If in difficulties, contact: Ordnance Survey, Department 32, Romsey Road, Maybush, Southampton SO9 4DH.

3.2.2 *North America*

The Geological Survey is responsible for publishing most of the topographic maps of the United States. Maps are published of the United States itself, Puerto Rico, Guam, American Samoa and the Virgin Islands at 1:1,000,000, 1:250,000 and, in certain areas only, at 1:62,500 and 1:24,000. Special maps are also printed at other scales. A free descriptive booklet can be obtained from: National Cartographic Information Center, Geological Survey, Reston, Virginia 22092. Maps can be bought from the Branch of Distribution, Geological Survey, at either Arlington, Virginia 22202 or the Federal Center, Denver, Colorado 80225.

In Canada the situation is more complex because both the federal and provincial governments produce maps. Enquiries should be made to the Map Distribution Office, 615 Booth Street, Ottawa 4. Maps likely to be used by field geologists are published at 1:250,000, 1:125,000 (in progress), 1:50,000 (in progress) and, in urban areas, 1:25,000.

3.2.3 *Australasia*

Australia produces maps at 1:1,000,000 and 1:250,000, with incomplete coverage at 1:100,000 and 1:50,000. They are available from: Map Sales Section of the Department of National Development, Tasman House, P.O. Box 850, Canberra City, ACT 2601; 460 Bourke Street, Melbourne, Victoria; and the Commonwealth Centre, Sydney, N.S.W. 2000. Maps and aerial photographs are also available from the Surveyor General, Department of Lands in each state or territorial capital, including both Tasmania and Papua.

New Zealand produces only two topographic map series of interest to the field geologist, namely one at 1:250,000 and an incomplete coverage at 1:63,360 (1 inch = 1 mile). Maps may be obtained from the Map Centre, Department of Lands and Survey, P.O. Box 6452, Te Aro, Wellington.

3.2.4 *Other countries*

Only general advice can be given on the availability of maps of other countries. Conditions vary greatly. Most countries of western Europe have good maps at 1:25,000 and 1:50,000. A few countries have 1:10,000 maps of limited areas and those of Switzerland are superb. There is little difficulty in getting maps of western Europe but those of eastern Europe and Asia are more difficult to come by. In many countries foreigners cannot obtain maps at all unless attached to a government department, and even then you must return them before leaving the country. Fortunately, in those countries, academic research is often tied to the mapping programmes of the official geological surveys, and this does ease the problem if you collaborate with their universities. In British Commonwealth and ex-Commonwealth countries there are usually good 1:50,000 topographic maps available but there are still many countries which have no maps suitable for geological purposes. In such places a geologist must work with a surveyor, or make his own base map by plane-tabling or from aerial photographs. An alternative is sometimes possible where good quality small-scale maps can be found, by photographically enlarging them to three or four times their original scale.

3.3 Geographic coordinates and metric grids

3.3.1 Geographic coordinates

Geographic coordinates represent the lines of latitude and longitude which subdivide the terrestrial globe. To make a map, part of the curved surface of the globe is projected onto a flat surface. This may result in one or both sets of coordinates being shown as curved lines, depending on the projection used. In Transverse Mercator's projection, however, the one most commonly used for the large-scale maps on which geologists work, latitude and longitude appear as interecting sets of straight parallel lines. This results in some distortion because, of course, lines of longitude in reality converge towards the poles, but on any single map sheet, the distortion is negligible. What does matter is that as latitude increases north or south of the equator, 1° of latitude remains almost (the earth is not a perfect sphere) a constant length of 60 nautical miles, but 1° of longitude becomes progressively shorter. Consequently, the use of geographic coordinates for pinpointing locations in the field, is to say the least, cumbersome.

3.3.2 Metric grids

The 'kilometre grid' printed on maps is a geometric, not a geodetic device. The grid is superimposed on the flat map projection and has (almost) no relationship to the surface of the globe. It is merely a system of rectangular metric coordinates, usually printed to give 1 km squares on maps of from 10,000 to 50,000 and 10 km squares on smaller scales. The grid covering Britain is numbered from an origin 90 km west of the Scilly Isles and extends 700 km eastwards and 1300 km to the north. For convenience, it is divided into 100 km square blocks, each designated by two reference letters. Other countries have other origins for their grids, and some use other systems.

The metric grid is a useful device for describing a location on a map. In Britain a full 'map reference' is given by first quoting the reference letters of the 100 km square block in which the point lies, e.g. 'SN' if in southwest Wales. This is followed by the *easting*, i.e. the distance in kilometres from the western margin of square SN, and then the *northing*, the distance from the southern margin of the square. The complete reference is written as a single group of letters and figures. For instance, SN8747 means that Llanwrtyd Wells is 87 km east and 47 km north of the margins of square SN. This reference is good enough to indicate a general area or a town. SN87724615, however, is more specific and locates to within 20 m the turn-off to Henfron from the Llanwrtyd Wells main road (Fig. 3.1), i.e., 87.720 km east and 46.150 km north of the square margins. These references are taken from the British 1:50,000 OS sheet No. 147. At larger

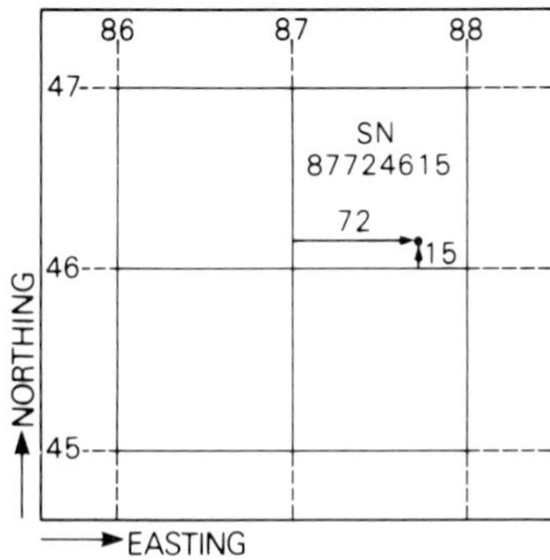

Fig. 3.1 Finding a map reference. The figure shows coordinates of a portion of 100 km square 'SN' of the British National Grid. The point referred to lies 0.72 km east of the 87 km coordinate and 0.15 km north of the 46 km coordinate. Eastings are always quoted before Northings.

scales, even more accurate references can be given.

Map (or grid) references are a convenient way of referring in a report or notebook to places on a map. They can designate areas, exposures, sample localities and geological observations. Geologists, also, usually plot their compass bearings from the grid lines on their maps rather than from lines of longitude, yet many still adjust their compasses to offset the difference between *magnetic* north and *true* north when they should be adjusting them for the difference between *magnetic* and *grid* north. In Britain, true north and grid north diverge to almost 5° in the Hebrides. Ensure that you adjust your compass against the proper variable (see Section 3.5).

3.4 Position finding on maps

In the field a geologist should be able to pinpoint himself to better than 1 mm of his correct position on the map, whatever the scale he is using; i.e. to within 10 m on the ground, or better, on a 1:10,000 map, and to within 25 m on a 1:25,000 sheet. On British 1:10,000 maps, a point may often be fixed merely by inspection, or by pacing along a compass bearing from a field corner, building or stream junction printed on the map, or by resecting from known points. If not, temporary cairns can be built to help him to locate himself. Where maps of poorer quality have to be used, a geologist may have to spend several days erecting points to work off and in surveying in their positions.

3.4.1 Pacing

Every geologist should know his pace length. With practice he should be able to pace with an error of under 3 m per 100 m over even moderately rough ground. This means that when using a 1:10,000 map he should be able to pace 300 m and still remain within the 1 mm allowable accuracy and to over half a kilometre if using a scale of 1:25,000.

Establish pace length by taping out 200 m over the average type of ground found in the field. Pace the distance twice in each direction and always count double paces, for they are less likely to be miscounted when pacing very long distances. Use a steady natural stride and on no account try to adjust it to a specific length, such as a yard or metre. Look straight ahead so that you do not alter your stride in the last few paces of each measurement to try to get the same number as last time.Every measurement should be within two double paces of the average of the four.

Prepare a table of paces, xerox it, and tape one copy into the back of your notebook and one into your map case (Table 3.1). When using this table, remember that you shorten pace when going both up and down hill so that allowances must be made to avoid overestimating. This is a matter of practice. If very long distances need pacing, pass a pebble from one hand to another, or from pocket to pocket, at the end of every 100 paces to save losing count.

Table 3.1 Example of a table of pace lengths in metres arranged for quick estimation of distances

Paces	*Metres*	*Paces*	*Metres*
1	1.7	10	16.6
2	3.3	20	33.3
3	5.0	30	50.0
4	6.6	40	66.4
5	8.3	50	83.0
6	10.0	60	100.0
7	11.6	70	116.6
8	13.3	80	133.2
9	15.0	90	150.0

3.4.2 Location by pacing and compass bearing

The simplest method of finding your position on a map—if mere inspection is insufficient—is to stand on the unknown point and measure the compass bearing to any nearby feature which can be identified on the map. Then pace the distance to the feature, providing it lies within the limits of allowable accuracy for the scale of map used. Plot the back-bearing from the feature, convert paces to metres, and mark the distance along the bearing with a scale.

3.4.3 Offsets

Offsetting is a simple method of plotting detail onto a map. It is particularly useful where a large number of points are to be plotted in one small area. Pace a line from a known position along a compass bearing until a point is reached directly opposite the first exposure to be examined. Drop your rucksack and then pace to the exposure at right angles to the main bearing line. This side line is an *offset*. Make your observations and then return to your rucksack and resume your 'traverse' along the same bearing as before until opposite the next exposure (Fig. 3.2). This method is comparatively fast for once the direction of the traverse or 'chain line' has been determined, preferably by lining in on a feature such as a tree, there is no real need to use your compass again except as a check; the right angles for the offsets can usually be estimated providing offsets are kept short. All you need to do is to count paces.

A variation of this method can be used on maps which show fences and walls. Pace the distance along a fence from a field corner and measure offsets from the fence-line to any exposure and observation points which need to be recorded. If the fence is long, take an occasional compass bearing to a distant point to check your position by 'intersection'. Students seldom make enough use of walls and fences although they are clearly marked on many maps.

3.4.4 Compass intersection

Your position on any lengthy feature marked on the map, such as a road, wall, fence, footpath, stream or river, can easily be found by taking a com-

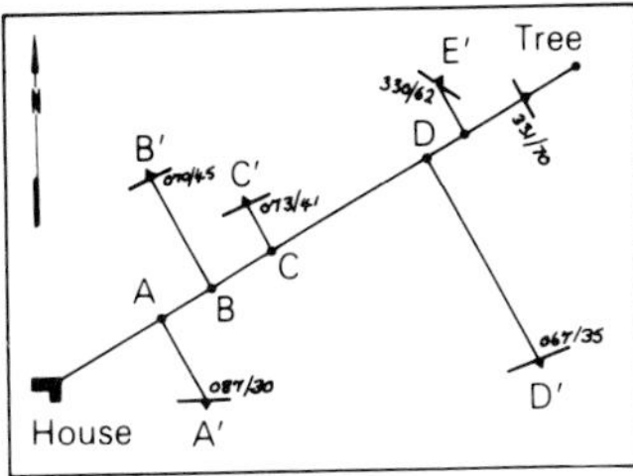

Fig. 3.2 Locating points by offsets. A traverse (bearing 62°) is paced from the house, using the tree as an aiming point, until you reach point A, directly opposite an exposure at A'. Mark A with your rucksack and then pace the offset A–A' at right angles to the traverse line. Plot the position of A' and make your observations. Return to your rucksack, resume pacing and repeat the procedure for B–B', C–C', etc.

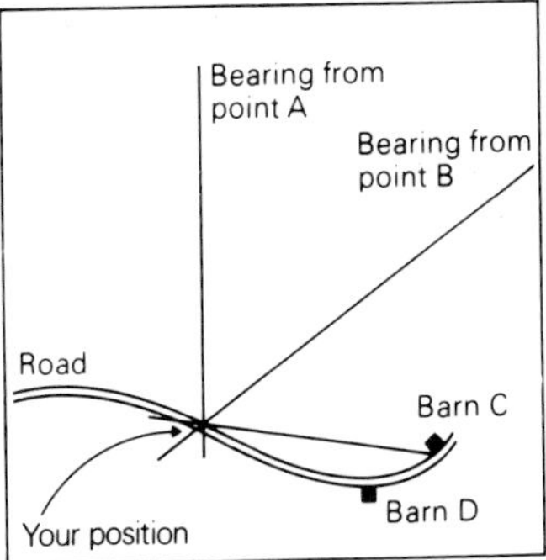

Fig. 3.3 Locating yourself on a road or similar longitudinal feature, sight points which give good intersections with the road; a bearing to the nearby barn, for instance, is not satisfactory.

pass bearing on to a point which can be identified on the map. Plot the back-bearing from this point and your position is where it intersects the road, wall, fence, etc. Check with a second bearing from another point. Choose your points so that bearings intersect your feature at angles between 60°–90° for the best results (Fig. 3.3).

3.4.5 Compass resection

Compass resection is used where ground is too rough or too steep, or distances too large, to pace. Compass bearings are taken from the unknown point to three easily recognizable features on the map, chosen so that back-bearings from them will intersect one another at angles of between 60°–90° whenever possible. Ideal intersections, unfortunately, are seldom possible, but every attempt should be made to approximate to them (Fig. 3.4). Features on which bearings may be taken include a field corner, a farmhouse, a sheep pen, a road, path or stream junction, a 'trig' point, or even a cairn that you yourself may have erected on a prominent point for this very purpose.

All too frequently, bearings do not intersect at a point but form a *triangle of error*. If the triangle is less than 1 mm across, take its centre as your correct position, if larger, check your bearings and your plotting. If the triangle still persists, sight a fourth point, if one can be found. If the error cannot be eliminated, it may be

Fig. 3.4 Intersection of bearings; (*a*) relatively good; (*b*) poor; (*c*) shows a triangle of error.

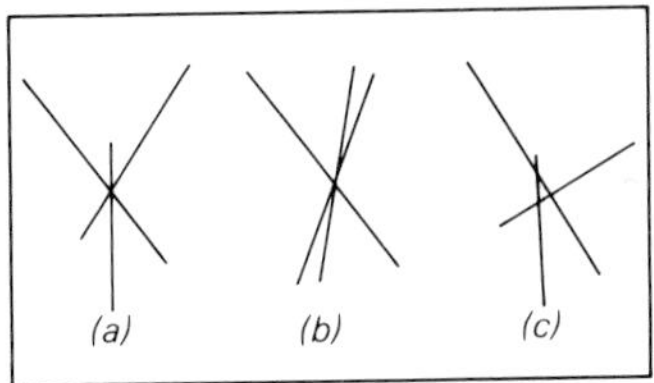

because the correction for magnetic declination has been wrongly set on your compass, or you may be standing on a magnetite-bearing rock such as serpentine, or you may be too close to an iron gate or under a power line. You may even have read your compass with your hammer dangling from a loop round your wrist—yes, it has been seen to be done! Or perhaps your compass is just not good enough for the job.

3.4.6 Compass and hand-level intersections

Where there is a lack of points on which compass bearings can be taken, a hand-level can be most useful. This is a device which allows you to sight a truly horizontal line. Such devices are built into the Brunton and Meridian geologist's compasses, while the Abney level is specifically designed as a hand-level. Some clinometers can also be used in the same way. To find your position, establish the contour you are standing on by scanning the surrounding hilltops, ridges and saddles with the hand-level until one is found at your own elevation. Providing you can find a feature less than 1 km away within ½° of your own level, you should be able to determine your contour to better than 10 m. Your position can then be established by a back-bearing from any point which will give a good intersection with your contour. Although not precise, the method may be all you can do in some places (Fig. 3.5).

Fig. 3.5 Levelling in a contour by hand-level. Set the level to zero and then search for a feature within ½° of your level-line.

3.4.7 Compass and altimeter intersections

An altimeter is an aneroid barometer equipped with an adjustable additional scale graduated in altitude above sea level. If set to read the correct altitude at the start of a traverse, and providing the barometric pressure remains constant, the altimeter should show the true elevation wherever you go that day. Unfortunately, barometric pressure is not constant. It has a regular variation throughout the day—the diurnal variation—and superimposed on that are more erratic variations caused by weather. The usefulness of altimeters is probably undervalued owing to geologists who have become exasperated in trying to use them in the wrong conditions for the wrong purposes.

Use them in a similar way to a hand-level, that is, establish the contour on which you are standing so that a simple compass intersection will then determine your position on that contour. The method is particularly useful in featureless country. The main problem is the variation in barometric pressure. This can be controlled in several ways. In very stable conditions spend a day in camp recording pressure changes on a graph which can then be used in the field to correct diurnal variation. In the field, check your barometer every time you occupy a point of known elevation on the map. If altitude readings are only occasionally needed, read your altimeter when you reach a point you cannot locate by other methods, and find the difference in altitude by returning to a point on a known contour. Better still, go back to a known elevation, return to the unknown, and then continue to another known height. You can then correct for any changes in pressure between readings.

3.4.8 *Siting additional survey points*

Temporary survey points can be erected to aid position finding, especially when working in a valley where it is difficult to see hilltop features. Build cairns of stones on higher slopes and survey them in by resection from well-established points. If wood is cheap, tall, flagged poles can be used in place of cairns.

3.5 Magnetic declination

At most places on the surface of the earth there is a difference between the direction of true north and the north shown by a magnetic compass. This is called *magnetic declination* or *magnetic variation* and it changes by a small amount every year. Magnetic variation, and its annual change, varies from place to place, and these values, together with the difference between true and grid north (which is, of course, constant), are shown as part of the marginal information printed on a map sheet. In Britain it amounts to about 1° every 15 years.

Magnetic declination must be allowed for when plotting compass bearings. As in most instances bearings will be plotted on the map from a grid coordinate, the correction used must be the difference between magnetic and grid north, not between magnetic and true north (see Section 3.3.2). On many needle compasses, such as the Silva and Brunton, this correction can be compensated for by rotating the graduated ring by means of a small screw. The compass will from then on give its readings in relation to grid north. Card compasses cannot be compensated: they can show only magnetic bearings and so every reading taken must be corrected. With practice, you do this in your head without thinking about it.

Many people prefer to establish their own correction by taking a bearing between two points on the map, or along a long straight feature, such as a moorland fence or wall, and then comparing it with the bearing measured on the map itself. This satisfies the doubter that he is not subtracting a correction that should be added, or vice versa.

3.6 Planetable mapping

Planetabling is a method of constructing a map for which little training is needed. It is excellent for making a geological map when no topographic base is available. In the first instance, the map, both topographic and geological, is made in the field at one and the same time. The contours are drawn with the ground in front of you so you can show all those subtle changes in topography which often have geological significance, differences which surveyors cannot show on tacheometer surveys where contours are drawn from spot heights plotted back in the office. Second, the plan position and elevation of every geological observation is accurate because it has been surveyed in. There is considerable satisfaction in plane-tabling, too, for your map grows before your eyes as geological and topographic detail is added. Plane-tabling makes you wholly independent of base maps of dubious quality, or of the assistance of topographic surveyors, who are not always available. It is described in many books on field geology, such as Reedman (1979), and Compton (1966), and any textbook on surveying.

3.7 Aerial photographs

The value of aerial photographs to the geologist cannot be overestimated. In reconnaissance, large tracts can be mapped quickly with only a minimum of work done on the ground. In more detailed investigations, examination of photographs under a stereoscope can reveal many structures which are difficult to see in the field, and some which cannot be seen at all at ground level. Photographs are as much a tool for the field geologist as his hammer and handlens. Good base maps do not obviate the need for photographs; they should be used together.

Aerial photographs can also be used where no base maps are available by building an 'uncontrolled mosaic' as a substitute on which geology can be plotted. It is not an accurate map, but it will serve its purpose for want of anything else. Information may also be plotted in the field directly onto photographs and then transferred to a base map later. This is particularly useful when the topographic detail on the base map is so poor that position-finding is difficult and time-consuming in the field. Excellent topographic maps can be made from photographs by a number of different techniques, but this is beyond our present scope.

Fig. 3.6 is a diagram of a typical aerial photograph. As each exposure is made, a photograph of a clock, altimeter, compass and circular level is also recorded in the *title strip* at the bottom of the photograph, to show time, height and tilt. The strip also shows the contract number, sortie number, and often, either the nominal scale of the photo or the focal length of the camera lens. Each exposure is numbered. *Fiducial marks* are printed at the corners or midway along each side of each photograph so that the *principal point* (see Section 3.7.1) can be marked on it if the camera does not print it on automatically. Different

Fig. 3.6 Layout of a typical aerial photograph, showing the fiducial marks at corners and mid-points of sides, the principal point, the title strip at the bottom, and the photograph number in the right-hand corner.

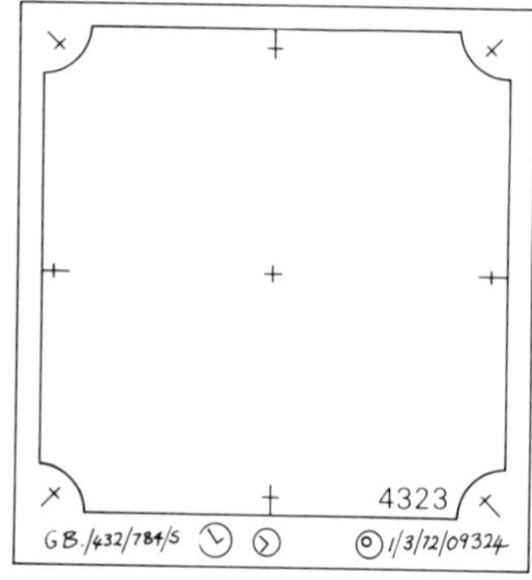

makes of camera have different title strip information, and arrange it differently.

Aerial photographs are taken sequentially by an aircraft flying along a series of parallel flight paths which may be straight lines or arcs of circles, depending on the method of controlling direction. They are taken with a frequency such that each photograph on a flight line overlaps the next by 60%, and each line of photographs overlaps the next by 30%. This apparently wasteful overlap is so that adjacent photographs on a line can be viewed under a stereoscope to produce a three-dimensional image, and also to ensure that there are enough common points on photographs to link them together for topographic map-making (Fig. 3.7).

Because the scale of an aerial photograph is a function of the focal length of the camera lens divided by the height of the camera above the ground, the true scale on an aerial photograph varies from place to place. A hilltop is closer to the camera than a valley bottom, and the centre of the photo closer than a corner of the photo. These differences cause distortions (Fig. 3.8). Distortion can, however, be removed to produce true-to-scale 'orthoprints'.

3.7.1 *Preparation*

Before they can be used, aerial photographs must be *base lined*. First mark the *principal point* (*pp*) on each consecutive photograph: this is the point where the optical axis of the lens meets the negative (Fig. 3.6). Now, locate the position of the *pp* (a_1 on Fig. 3.7) of your first photograph on the overlapping part of the second

Fig. 3.7 A block of three runs of aerial photographs, A, B, and C. Photographs in each run overlap by 60% so that the position on the ground of the principal point (a_1) of photograph A–1 can also be found on photograph A–2. Similarly the *pp* a_2 on photo A–2 is found on both photographs A–1 and A–3. Adjacent runs overlap by about 30% so that the feature d_1, seen on photograph B–1, can not only be found on photographs B–2 and B–3, but also on photographs C–1, C–2 and C–3 of the adjacent run.

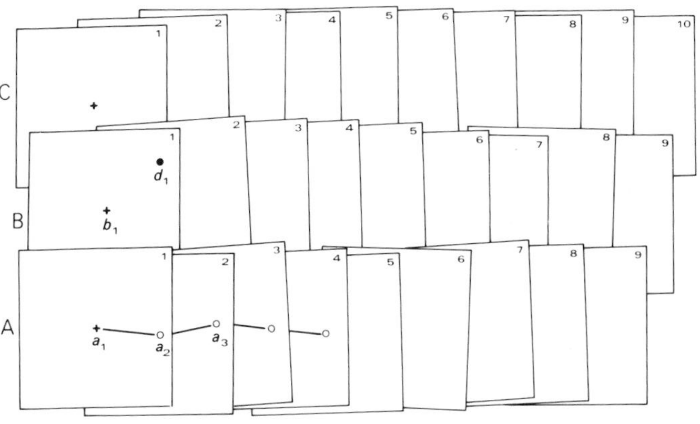

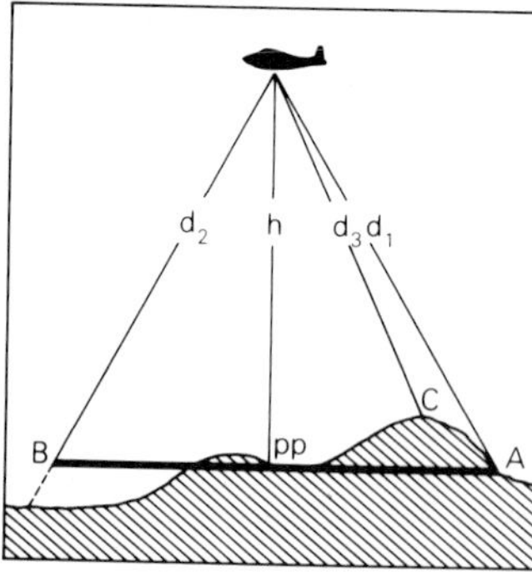

Fig. 3.8 Scale variations in an aerial photograph taken in undulating country. A–B represents the notional plane of the photograph; *pp* is its principal point. The nominal scale of the photograph is *f*/*h*, where *f* is the focal length of the camera lens and *h* is the height above the ground at *pp*, i.e., its distance from the lens. At A, the distance from the lens (d_1) is greater than *h*, therefore the scale of the photograph there is smaller than at its centre, *pp*. At B, the fall of the land means that d_2 is even greater than d_1 and consequently, the scale is smaller still. The distance d_3 to hilltop C, however, is less than *h*; the scale on high ground is therefore less than at *pp*.

photograph in the run. Do this by inspection with a handlens and prick its exact position on photograph No. 2 with a needle. Draw a small circle round the needle hole. This transferred principal point is called a *conjugate point* (*cj*). Now transfer the *pp* of photograph No. 2 to photograph No. 1 (point a_2), and to photograph No. 3, and so on. Draw base lines between the *pp* and the two *cjs* on every photograph (Fig. 3.7). Base lines indicate the track of the aircraft between photographs and show that the flight line is seldom straight owing to drifting and yawing. The purpose of the base line here, however, is to make it easier to align your stereopairs of photographs under the stereoscope in their optimum position to give a 3-dimensional image. Although your eyes will accommodate the strain needed to produce a 3-D image from poorly positioned photographs you will end the day with a headache.

3.7.2 *Plotting on aerial photographs*

The surface of the photograph is not easy to write on in the field. The best method of recording information is on an overlay of tracing film, such as 'Permatrace' or 'Mylar'. Cut the overlay to the same size as the photograph and tape it on along *one* edge only by a hinge of *drafting* tape, so that it can be lifted whenever you wish to examine the photograph more clearly. Do not use transparent tape as it damages the photograph when it is removed. Mark the photograph number and the *principal* and *conjugate points* on the overlay, so that it can be repositioned on the photograph if it is necessary to do so at some later date.

Locating your position on a photograph is usually easy. It can be done either by inspection of a single photograph to identify a nearby feature, or if in difficulty, the 'stereopair' can be used with a pocket stereoscope to give a 3-D image of the ground. Note that the three-dimensional image seen under a stereoscope gives very considerable vertical exaggeration to the topography. Small hills look like high hills, high hills look like saw-tooth mountains, and this exaggeration must be taken into account in locating yourself. What you cannot do on a photograph is locate yourself by compass resection.

3.7.3 *Northpoints on photographs*

Structural information is plotted onto a photograph in the same way as onto a base map. Photographs, however, are seldom taken along flight lines which run north–south and even if they were, aircraft yaw sufficiently that the margin of the photograph could still not be taken as the intended flight direction. Therefore a northpoint must be established for every individual photograph on every flight line. This can only be done on the ground. Position yourself as close to the centre of the photograph (the *principal point*) as possible and take a compass bearing on some easily identifiable feature on a line as nearly radial from the photo-centre as possible. This is because, owing to linear (i.e., scale) distortions, the only true bearings between points on a photograph are those which originate from the principal point. This does not affect the plotting of strikes and lineations at single points on a photograph.

3.7.4 *Transferring geology from photograph to base map*

Geology plotted onto photographs or overlays in the field must be transferred to a base map later. You cannot trace information directly from one to the other because they will never be exactly the same scale. *Camera lucidas* are available which enable the map to be viewed with an image of the photograph adjusted to the same scale superimposed upon it. Information can also be transferred directly from photographs by inspection. Dips and strikes must then be replotted from recorded compass bearings. Because of a lack of fine topographic map detail, however, it may not always be possible to locate on the map an observation point marked on the photograph. In that case, draw a radial line from the principal point on the photograph to the observation, measure the angle it makes with the northpoint on the photograph; then plot the same bearing from the position of the principal point marked on the map. The observation lies along this line and its exact position can usually be found from other information. If not, establish the difference in scale between map and photograph and plot the distance from the *pp* proportionately. This will not work in mountainous areas. In that event, measure the angle between the radial line and the *base line*; locate the same observation point on the adjacent photograph and measure the angle between the base line and that radial line; plot the angles from the ends of the same base line marked on your map. The point lies at the intersection of the radial lines (Fig. 3.9). This is 'radial line plotting'.

Before transferring any information from photographs by any method, the principal points of every photograph must be marked on the map. Some maps may even have the *pp*'s of the photographs from which they were compiled already printed on them, with the photo-numbers shown in very fine type beside them.

3.7.5 *Sources of aerial photographs*

In Britain photographs can be bought through the Central Register of Air Photographs of England and Wales, London; the Welsh Office in Cardiff; or the Scottish Development Depart-

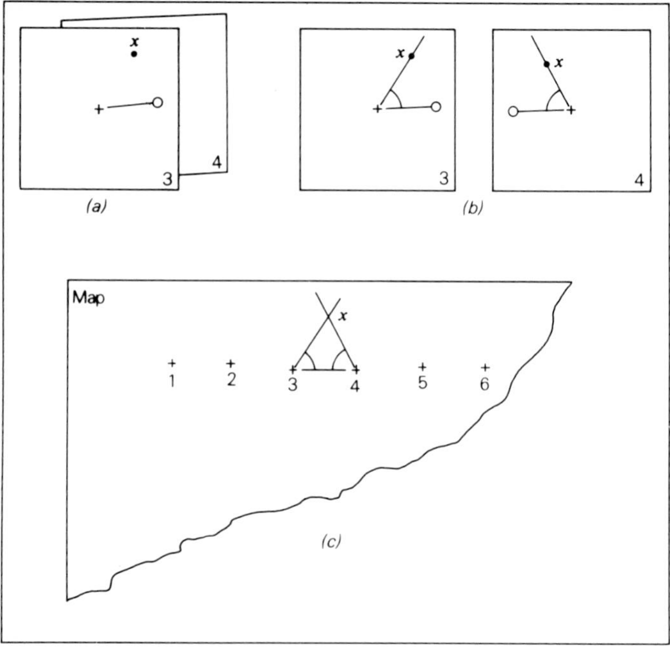

Fig. 3.9 To locate on a map the position of a point *x* seen on aerial photographs:
(a) Locate the position of *x* seen on photograph 3 on the overlapping photograph No. 4.
(b) Draw a radial line from the principal point of each photograph through the point *x*. Measure the angles made with the base lines.
(c) Join *pp*'s 3 and 4 on your base map; plot the angles measured from the photographs; *x* lies at their intersection.

ment in Edinburgh. They will tell you what photographs are available and at what scales, provided you give them the map references of the corners of the area you need cover for. Private contractors listed in the *Geologist's Directory* (Institute of Geologists 1980) may also have photographs for sale. In the USA photographs can be supplied by the Geological Survey, EROS Data Center, Sioux Falls, S. Dakota 57198. Sources in Australasia are given in Section 3.2.3. Elsewhere, photographs may be more difficult to obtain. They are often under direct military control and may be restricted even for a country's own nationals. As with maps, photographs may sometimes be obtained if you are attached to a government organization or a university of the country concerned.

4 Methods of geological mapping

Geological mapping is the process of making observations of geology in the field and recording them so that one of the several different types of map described in Chapter 2 can be produced. The information recorded must be factual, based on objective examination of rocks and exposures, and made with an open mind. Geology is too unpredictable to be appoached with preconceived ideas. Obviously, the thoroughness with which a region can be studied depends upon the type of mapping on which you are engaged. A reconnaissance map is based on fewer observations than, say, a regional map, but those observations must be just as thorough. Whatever the type of mapping, whatever your prior knowledge of an area, map with equal care and objectivity.

Although in this chapter distinctions may still be made at times between reconnaissance, regional and detailed mapping, the methods described are not exclusive to any one of them. Methods must be suited to the conditions, and varied as occasion requires. If none of the methods suggested here are adequate for a particular task, then develop new ones of your own.

4.1 Traversing

Traversing is basically a method of keeping track of your progress across country. Traverses are made by walking a more or less predetermined route from one point on the map to another, plotting the geology along the way. They should be planned to cross the general geological grain of the region. Traversing is an excellent way of controlling the density of your observations. In reconnaissance work a number of roughly parallel traverses may be run across country at widely-spaced intervals. Contacts and other geological features are interpreted between them. This leads to few complications in regions where the rocks are only moderately folded and dip faults are few, but reliability decreases as structure becomes more complex. Traverses can also be used to map in detail areas where rocks are well exposed, especially those where there is almost total exposure. In these cases, traverses are closely spaced.

In open country, where visibility is good and the base map adequate, traverse 'legs' may be walked from geographic feature to geographic feature; for instance from a hilltop to a bend in the river, or, on a larger scale, from farmhouse to field corner. Each leg should be a straight line, and the hilltop, farmhouse and field corners where the traverse changes direction are called 'turning points'. In forests, thick bush, broken topography, or where the base map is poor or non-existent, direction must be controlled by compass bearings. If the base map is good, the positions of geological observations on a leg are estimated from features or by compass resection, or by pacing, pedometer or cyclometer wheel, depending on the accuracy required. Plot geology directly onto the map as the traverse leg is walked. Distinguish geological fact from inference by showing a solid traverse line where a formation is well exposed, and a broken line where a rock can only be inferred. Overlay the lines in the appropriate coloured pencil for the rock type seen or inferred.

Before starting a traverse, examine the ground ahead, with binoculars if necessary, so that the best route can be chosen, both in terms of geology and accessibility. Aerial photographs can help you to plan your route too. Mark the end of each leg in case you need to return to check geology or correct a mistake in measurement. Use a stone, blazed tree, stake, or even a small cairn and number it with a felt-tipped pen or timber crayon.

Much emphasis has been placed here on traversing. This is intentional. Only too often geologists wander aimlessly from rock to rock, keeping little track of their movements. Every time they stop to make an observation they must relocate themselves from scratch—some, possibly, just making a guess. Traversing ensures you do cover the ground properly with the least wastage of energy and time by not having to continually search your map to find out where you are.

4.1.1 Controlling traverses

Unless traverses are strictly controlled, survey errors accumulate to an unacceptable level. Wherever possible, make your traverses from known point to known point. If a traverse consists of a number of legs controlled by compass bearings, start at one known point on the map and finish at another: alternatively, make a complete loop and finish back at your starting point. Invariably, you will find that the last bearing and distance plotted does not fall exactly where it should owing to the accumulation of minor errors due to the limitations of the measuring methods used. This *closure error* must be corrected by distributing it over the whole traverse: a convenient method is described in Appendix II.

Because a compass traverse always needs to be corrected, do not record geology directly on to the uncorrected traverse on your field map. Plot the traverse lines, from turning point to turning point, on your map, but record the details of the geology in your notebook as a sketch on an exaggerated scale. If your notebook is a surveyor's 'chain book' with a double red line down the centre of the page, then borrow the surveyor's technique. Use this column as if it were your traverse line. Record the distance of each observation from the start of a leg within this column and show the geology to either side of it (Fig. 4.1). This keeps distances measured along the traverse line and the details of geology separate.

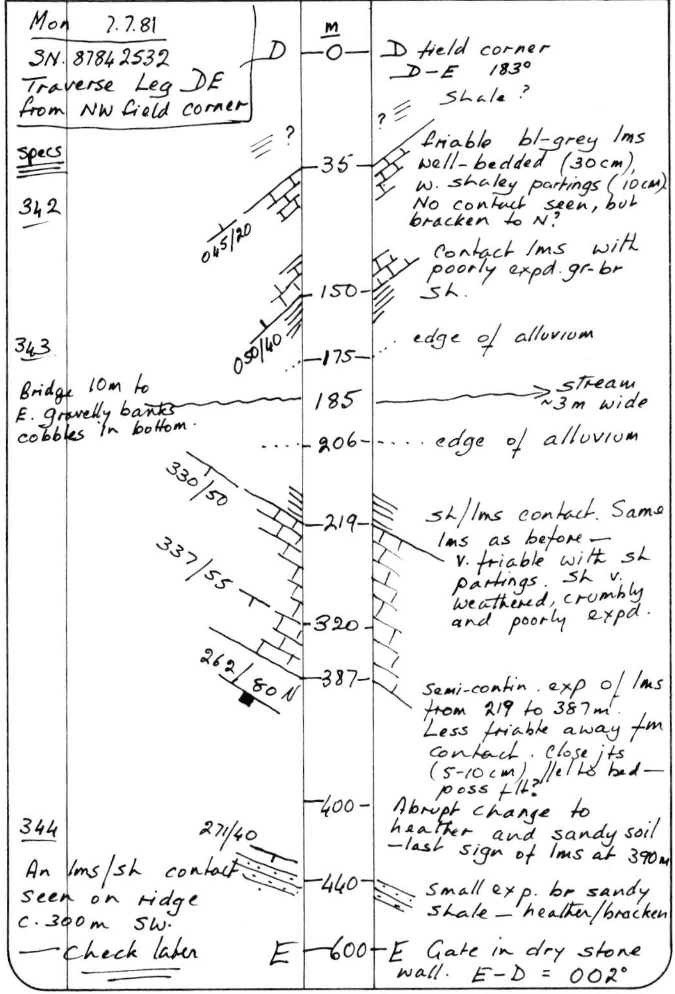

Fig. 4.1 Recording a traverse in a surveyor's chain book. The double column down the centre of the page (often printed in red) represents the traverse or 'chain' line. It has no actual width on the ground: it is used merely for recording the distance from the start of the traverse leg.

Wherever possible, correct your closure error in the field. When corrected, plot the geology on the map.

4.1.2 Cross-section traverses

Where succession is doubtful or structure complex, traverse across the geological grain, plotting a cross-section as you go. Draw it on squared paper kept for the purpose in your map case, or in your field notebook, but plot the traverse legs on your field map. The advantages of drawing sections in the field are obvious. Problems come to light immediately and can be promptly investigated.

4.1.3 Stream and ridge traverses

Streams and ridges are features which are usually identifiable on even poor quality maps. Streams often give excellent semi-continuous exposures and in some mountain areas may be so well spaced that a major part of the geology can be mapped by traversing them alone. Much reconnaissance work is based on stream traversing. Position finding on streams is often relatively easy from the shape and direction of bends, and the position of islands and other features and, if the surrounding country is open, compass bearings can be taken on distant points. In dense mountain rain forests, streams and rivers may, in fact, be the only places where you can locate yourself on your map or photographs because no other features can be seen through the canopy of trees.

Ridges, and the spurs which lead off them, may also make excellent traverse locations. They can usually be identified easily on a map or aerial photograph. Even in dense forest, ridges may be relatively open, giving opportunity to take bearings to distant points from them. Exposures are usually good. Most ridges are there because they are erosion-resistant, and in sedimentary rocks tend to follow the strike. Side traverses, down spurs, provide information on the rocks which lie stratigraphically above and below those of the crest; alternatively, stream traverses may provide better information.

4.1.4 Road traverses

A rapid reconnaissance of an unmapped area can often be made by mapping the geology along tracks and roads and by following paths between them. Roads in mountainous regions, in particular, usually exhibit excellent and sometimes almost continuous exposures in cuttings. In some places the roads zigzag down mountainsides to repeat exposures at several different stratigraphic levels. A rapid traverse of all roads is an excellent way of introducing yourself to any new area you intend to map in detail.

4.2 Following contacts

A primary object of mapping geology is to trace contacts between formations and show where they occur on a map. One way of doing this is to find a contact and to follow it on the ground as far as it is possible to do so. In some regions, and with some types of geology, this is easy; elsewhere it is often impossible because contacts are not continuously exposed. Following contacts is probably the easiest method of mapping but it is not al-

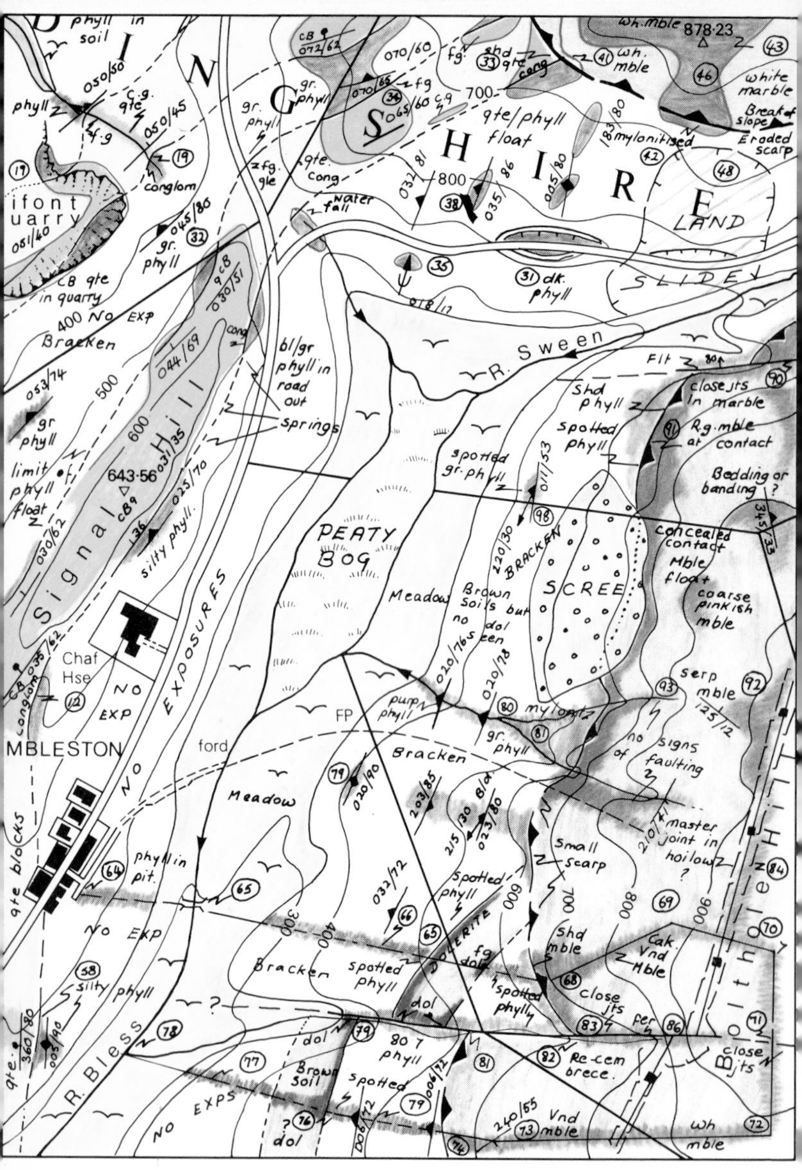

Plate 4.1 Portion of an inked-in field slip showing: green-line mapping in the north; exposure mapping without the green-line in the centre; and a closed traverse (points 64–78) and a stream traverse (points 78–84) in the south. Unexposed ground has been coloured in to distinguish ground already covered from that yet to be mapped.

ways the most rewarding. Areas where it can be used as the main mapping technique are usually structurally uninteresting. Sometimes contacts identified on the ground can be followed more easily—and more accurately—on aerial photographs under a stereoscope. The photographs show small changes in topography and vegetation which cannot be detected on the ground, but which indicate the position of the contact even where it is concealed by colluvium or other drift. Once traced on the photograph, the contact should be checked in the field at its more accessible points.

Wherever rocks are seen in contact, show the boundary as a continuous line on the map and mark each side with the coloured pencils appropriate to those rocks. Where contacts are inferred, show the boundary with a dashed line. Where a contact is concealed—for instance where it passes beneath alluvium—show it as a dotted line.

4.3 'Exposure' and 'green-line' mapping

Mapping by exposures is the mainstay of much detailed mapping at scales of 1:10,000 and larger. The extent of each exposure or group of exposures, is indicated on the field map by colouring it in with the appropriate coloured pencil for that formation. Some geologists go further and mark the limits of exposure by drawing a line round it, later inked in green—hence 'green-line' mapping. Green fades rapidly in the tropics and a fine black dotted line can be substituted. Whether or not you draw a line around your exposures is a matter of choice, but if a map is used in the field over long periods, exposures become blurred as pencil shading fades, or is worn off. If boundaries are inked, colouring can be touched up when needed; if not, exposure edges become vague and accurate recolouring difficult. Marking the boundaries of very large exposures helps objectivity in the field: outline the exposure, then map within it. If complex, or if there are specifically interesting features to be seen, a large-scale sketch map can be made of it in your notebook. Do not be too fastidious in plotting accurate outlines; an approximate shape is all that is required. On the other hand, the natural optimism of human nature nearly always results in exposures being shown larger than they really are unless some care is taken: remember that an exposure 10 m square is a mere 1 mm^2 on a 1:10,000 map; one the size of a football pitch is only 10 × 5 mm. Show groups of exposures which are obviously part of the same outcrop thinly covered with drift, as a single exposure. Mark small isolated exposures by a dot with a note or symbol to indicate its nature beside it.

The reason for exposure mapping should be clear. It shows the factual evidence on which your interpretation of the geology will be based; it shows what you have seen, not what you infer. A properly prepared field map should leave no doubt of the quality and quantity of the evidence it is based on. Plate 4.1 illustrates the general principles.

4.3.1 Descriptive map symbols

There are some types of geological terrain where the geology can be mapped only by identifying every exposure in turn; for instance, in Precambrian metamorphic terrains slates pass into phyllites, then to schists, migmatites and gneisses of several kinds. Many boundaries are gradational and contacts have to be decided by textural and mineralogical characteristics. In these conditions, the usual colour coding used to distinguish formations on you map is inadequate, although it may serve to classify your rocks into broad groups. You must devise a letter code so that you can give a shorthand description of every exposure on your map to show how metamorphic lithologies change and to decide your boundaries. You may need to distinguish, say, *microcline-porphyroblast coarse grained quartz-albite-microcline-muscovite-biotite gneiss* from other, not quite similar, gneisses. This could be condensed to *M c/gr q-ab-m-mubi gn*, where *M*, stands for microcline porphyroblast and *m* for microcline in the groundmass, etc. Devise your own code. These codes give *field-names* (Section 6.2.1), not to be confused with the *formation names* and letters (Sections 6.2.2 and 8.5) used to designate recognized formations. Try to keep them more concise than the extreme example quoted above. Whatever code you devise, make it flexible, for you will invariably find that you have covered only a proportion of the possibilities you may eventually meet in the field.

4.4 Mapping in poorly exposed regions

If an area is poorly exposed, or the rocks are hidden by vegetation, climb to convenient high ground and mark on your map the positions of all the exposures you can see. Then visit them. Of all rocks, mica schists probably form the poorest exposures but even they may show traces on footpaths where soil has been worn away by feet or by rainwash channelled down them. Evidence of unexposed rocks may also sometimes be found where trees have been uprooted by storms and in the spoil from holes dug for fenceposts or wells, in road and railway cuttings, and from many other man-made excavations.

4.4.1 Indications of rocks from soils

Soils, providing they are not transported, reflect the rocks beneath, but to a much lesser extent than expected. Sandy soils are obviously derived from rocks containing quartz, clayey soils from rocks whose constituents break down more completely. Dolerite (diabase) and other basic rocks tend to produce distinctive red-brown soils; more acidic igneous rocks form lighter-coloured soils in which mica may be visible, and often quartz. A soil depends not only on its parent rock, but also on climate and age. Differences tend to become blurred. When working in any area, poorly exposed or not, take notes wherever soils are seen to be associated with specific rocks so that they can be used as a guide when needed.

4.4.2 Topography and vegetation as a guide

Both topography and vegetation reflect geology and should be noted during mapping. Springs, seepage lines, lines of more luxuriant vegetation, and changes of vegetation, may indicate contacts, joints and faults, or a change of rock below. Some plants thrive on soils formed from certain rocks, but do not grow on others; take notes for future reference where plants and rocks can be correlated. Add topographic symbols to field slips to indicate minor features which are not already printed on them, but which might reflect underlying geology or structure. Breaks of slope, or small scarps and ridges which are too low to be reflected in the contours, are typical examples.

4.4.3 *Evidence from float*

Many soils, particularly on hillslopes, contain rock fragments called 'float'. Fragments from the more resistant rocks tend to be large and may lie on the surface. Those from softer rocks are smaller and usually buried; they have to be dug for with the sharp end of your hammer or with an entrenching tool. Contacts on hillsides can sometimes be located with considerable precision by searching for the upper limit of float derived from a formation which lies immediately below a contact with another rock (Fig. 4.2). Care must obviously be taken in glaciated regions that the hill-slope soils, or colluvium, have not been transported.

Fig. 4.2 Float in hillslope colluvium as an indicator of a contact. The contact is where the first signs of hard rock float appear in soil.

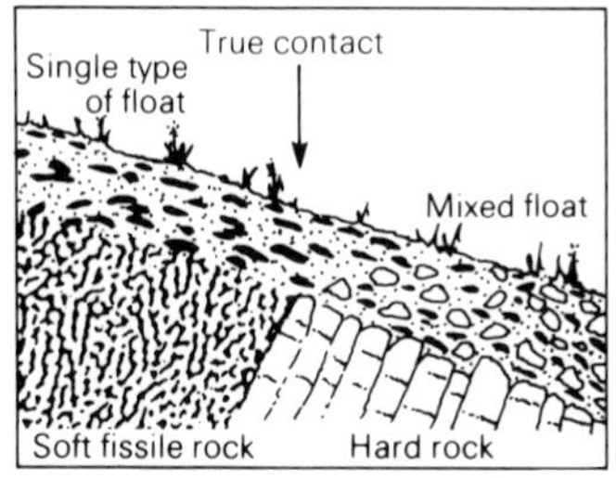

4.4.4 *Pitting, trenching, augering and loaming*

When it is essential to examine rock beneath the soil in a poorly exposed region, pits and trenches may have to be dug. A pit can be sunk quite rapidly as long as the digger does not become too ambitious over the size of the hole. The most economical pits, widely used for prospecting in Africa, are about 85 cm in diameter, excavated with a short-handled hoe. Contacts are best located by trenches (*costeans*).

In many cases, identifiable fragments of weathered rock can be recovered from shallow hand-auger holes. A 'post hole' auger can rapidly sink a 10–20 cm diameter hole to 60 cm. Mechanized augers are obviously quicker.

Loaming is a method of mapping in poorly exposed, deeply weathered, regions. Soils, collected from beneath the humus layer in pits and auger holes, are washed in a gold pan (Section 5.11) and the concentrates compared with 'heavy mineral suites' collected from soils lying above known formations. Large areas of Venezuela, and smaller, laterite-covered parts of Africa, have been mapped in this way.

4.5 Drilling

Every geologist will be concerned with drilling at some stage of his career. It is most commonly employed to locate formations at depth; to confirm their presence in the lack of other evidence; to solve structural problems; and to sample rocks and ores. It is also used to find and exploit water and, of course, oil.

Basically, there are two types of drill: percussion ('churn') and rotary. Percussion rigs drill by repeatedly raising a heavy drill bit attached to a wire cable and dropping it to just touch the base of the hole. Rock is *crushed* and *chipped* away and the debris is bailed from the hole at intervals for examination. Rotary drills, on the other hand, rotate a drill bit attached to the end of a tubular drill pipe: the rock is *ground* away. Frequently, but by no means always, the bit is set with diamonds—hence 'diamond drilling'. Some rotary bits are tubular and cut a ring-like hole which leaves a cylindrical 'core' of rock attached to the bottom of the hole. This can be broken off and recovered as a sample of solid rock. Percussion rigs can drill only vertical holes: they yield chippings without core from holes about 20 to 60 cm in diameter. Rotary rigs can drill inclined holes and they can, but do not have to, take a core. Holes can be from about 4 to 60 cm in diameter, with the larger holes drilled by 'tricone bits' consisting of three conical cutting wheels. The sludge of ground-up rock flour formed by rotary drilling is continually removed from the hole by circulating drilling fluid and collected from the hole as sample material, whether core is taken or not (Fig. 4.3).

Fig. 4.3 A diamond drill drilling an inclined hole.

4.6 Geophysical aids to mapping

Geophysics plays an increasingly important role in geological investigations and every geologist should know how it can be applied so that he can ask for appropriate help when needed. Most geophysical methods need a specialist geophysicist to apply and interpret them, but there are a few instruments that a geologist can use himself to help him to locate concealed contacts. They are available in most geological organizations and two are described below.

4.6.1 Magnetometers

Compact torsion-balance magnetometers are available, small enough to be operated in the hand. They are adequate for distinguishing between rocks with no magnetite and those with magnetite. For example, they can find the contact between serpentine and the surrounding sedimentary rocks or locate unexposed dolerite (diabase) dykes. Less portable, but more sensitive, are *proton precession* instruments.

4.6.2 Radiometry

Acid igneous rocks, rich in potash feldspar, contain sufficient ^{40}K (potassium-40) to enable them to be distinguished from rocks with lesser K-feldspar nearby if a sufficiently sensitive instrument is used and the soil cover thin. A gamma-ray spectrometer (*scintillometer*) will detect these differences although the older *Geiger counter* cannot.

4.7 Superficial deposits

Only too often, unconsolidated deposits are mapped poorly, if at all. Superficial deposits, or 'drift', must be indicated on field maps. Laterites, sand dunes, boulder clay, and river and beach deposits represent important events in the later geological history of a region. Peat and areas of bog and swamp may indicate recently past climates, or that drainage has been disturbed, perhaps by tilting or other causes. Unconsolidated materials of lesser extent, such as scree, landslide debris and hillside colluvium should be noted too. Nor should soils be ignored. All, apart from any other consideration, obscure the solid geology and their presence—or absence—contributes to the reliability of your interpretation. Much superficial material is clearly visible on aerial photographs and can be plotted on to your field map directly from them.

Not all this information need be transferred to your fair copy map. Soil and colluvium, for instance, justify your interpretation to those who refer to your field map later but add nothing to the understanding of the geology of the region. Scree need be transferred only if it is extensive or covers important contacts so that interpretation beneath it is speculative. Much depends on the map scale. Scale particularly affects the mapping of alluvium: at small scales it can be generalized but on more detailed maps individual terraces may be shown. Dunes, laterite and boulder clay, however, are part of the stratigraphic succession and must always be shown.

4.7.1. Landslides

Landslides are a special case. They are a geological hazard and far more common than most geological maps would suggest. To ignore them is geological negligence. If not recognized, much time can be wasted in trying to make structural sense from the diverse strikes and dips that sliding produces. That an apparent rock outcrop is the size of a house is no guarantee that it is in place. Evidence of sliding is important to both environmentalists and engineers who may use your map for planning. Builders of dams, roads, railways and housing estates also, not unnaturally, wish to know of unstable areas.

Landslides can be recognized by the scar where the slide starts, and by the material that has slid (Fig. 4.4). If the slide is old the scar may be eroded and overgrown. The debris, however, may show several recognizable features. Its average gradient is gentler than the rest of the hillside and its surface different. There may be small parallel ridges or hummocks

Fig. 4.4 A major landslide near Livingstone, Montana, USA. Note the hummocky nature of the slipped ground in front of the stable slopes forming the hills on the sky-line.

caused by 'earth flow'. Drainage is small scale, often dendritic, and there may be small ponds and pools. In heavily-wooded areas the slide may support only scrubby bush, or dead trees with new growth between them. Where sliding is imminent, trees may be 'kneed'. Some slides are indicated by massive unweathered blocks poking through the hillside colluvium and they can cover huge areas. Map a slide as a distinctive geological unit, indicating both the scar and the spread of the debris.

4.8 Large-scale maps of limited areas

From time to time there is the need to map specific aspects of the geology on a far larger scale than that being used for your main map. In Britain, it may be possible to use 1:1250 or 1:2500 OS plans, or prints from an OS *copy-card*. More satisfactory for those able to do it, is to use a planetable. This gives great flexibility in scale and accurate geological maps as large as 1:500 can be made this way. Even if no great accuracy is required, plane-tabling is often the easier way of making a large-scale map. It is certainly the best where the ground is rugged, broken or uneven, and wherever the correct vertical position of a point is as important as its plan position.

More often, the need arises for a very large-scale sketch map of a very limited area, sometimes only a few hundred square metres in extent. The need is to illustrate geology and no great precision is required. Thus, methods can be used which might well be derided by a land surveyor. Some are described below: they can be modified and changed to meet contingencies. Ingenuity and a basic knowledge of surveying are assets. Keep sheets of squared paper in your map case in case you need them.

4.8.1 Compass and tape traverse

The simplest method of plotting geological detail is by taking offsets from a 'chain line' or traverse, as described at Section 3.4.3. A single traverse

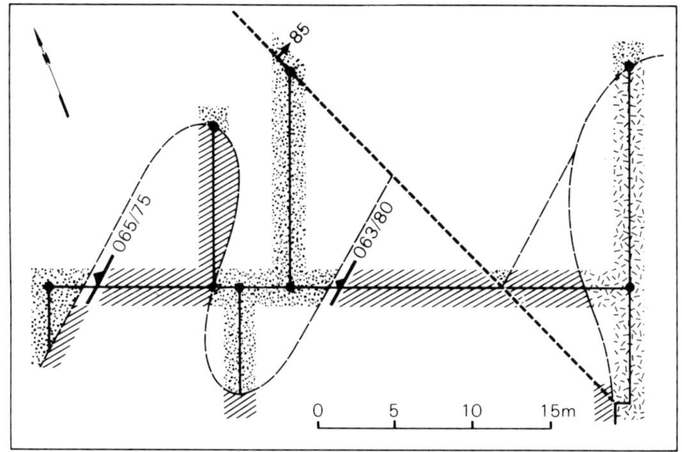

Fig. 4.5 Simple compass and tape traverse to plot in large-scale geological detail.

may even suffice (Fig. 4.5). The same method can be used as a 'mini traverse' to map a single exposure in detail.

4.8.2 Traverses with offsets

Where a number of exposures are spread over an area of more or less level ground, but scattered too far apart to be mapped by a single traverse line, geology can be mapped rapidly by running a series of traverse legs in a loop. Detail is mapped by offsets from the legs (Fig. 4.6). For small areas, measure the bearings and lengths of the legs first, marking the turning points so they can easily be found again. Plot the traverse and correct the closure error, then plot the geological detail. An alternative is to enter all details, including the geology, in your notebook as you move along each leg in turn, and re-plot everything back in camp. The first is to be preferred because then you have the ground in front of you as you plot the detail.

Fig. 4.6 A closed traverse of several legs to plot in a number of exposures for a medium-scale sketch map.

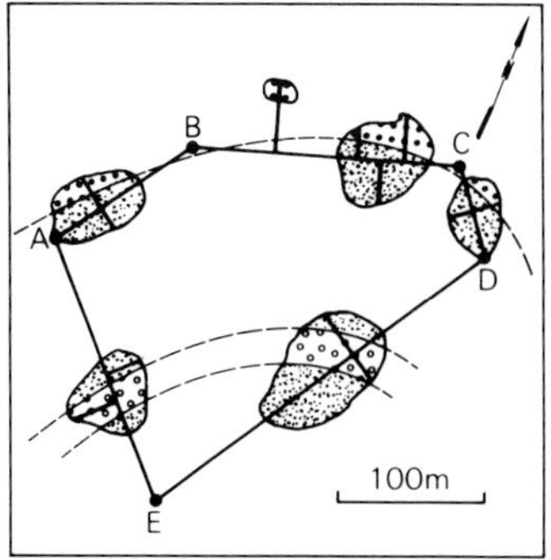

4.8.3 Mapping an exposure in detail

It is sometimes necessary to map a large area of exposure in detail. If the surface is more or less flat, lay down a base line; use stones to mark points along it at fixed intervals (say 10 m); and then measure traverses at right angles from it, with stones again marking 10 m intervals. The effect is to build up a grid to guide your sketch map (Fig. 4.7). Where a great deal of field sketching of exposures is to be done, a cord grid which can be laid over an exposure and anchored there with stones will simplify the task. The grid shown in Fig. 4.8 was constructed by pegging out an area 16 m × 20 m, with pegs every 4 m along the sides. Three-ply nylon cord was used to make a net with a 4 m mesh. Detail is plotted by estimation on squared, water-resistant paper, with measurements by steel tape when necessary. Compass bearings are measured by assuming one side of the net is 'grid north' and correcting your compass to read accordingly. Fig. 4.9 shows structure mapped in the deformed 'Scourie' dyke shown in Fig. 4.8.

Remember that other methods not described here may be used by other people. Use the methods that suit you and the geology best but ensure that those you do use give acceptable results in terms of accuracy. Different types of geological environment will affect the way you map, so will different terrains, different climates, and different base maps. Adapt and invent: never become hidebound.

Fig. 4.7 Mapping a large exposed area by building up a rough grid.

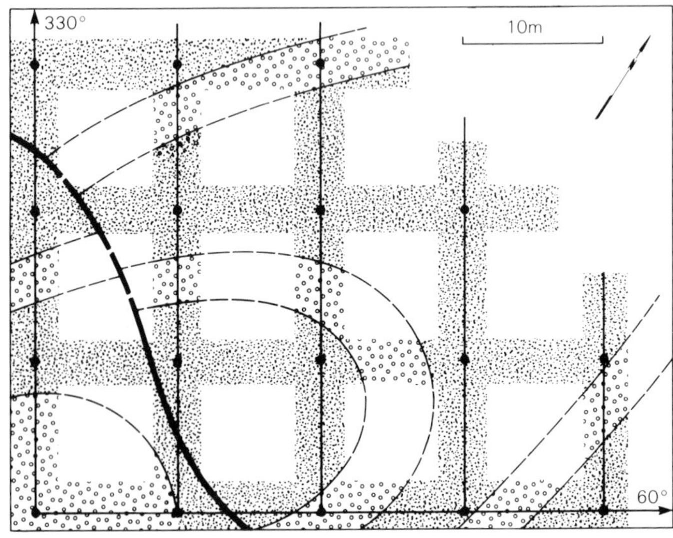

Fig. 4.8 A 4 m mesh cord grid to aid sketch mapping laid over an exposure in NW Scotland (see Fig. 4.9). (The cord in the photograph has been retouched for clarity.)

4.9 Photogeology

Photogeology is the systematic interpretation of geology from aerial photographs. It can be used as a method of geological reconnaissance with only limited groundchecking, or as an adjunct to orthodox geological mapping. Here, we consider only this second use.

4.9.1 Use of aerial photographs

Before leaving for your field area, examine your photographs under a mirror stereoscope and make an interpretation of the main geological features. When you reach the field, carry the photographs in your map case in addition to your field map. Examine them at intervals with a pocket stereoscope to compare what you see on the ground with its appearance on the photographs. At night, review your map and photographs together. You may well find that you can trace contacts and faults on your photographs which you could not trace on the ground. This is because the vertical exaggeration of the 3-D image seen under the stereoscope accentuates quite minor features which reflect geology. Check in the field next day to see if you can now locate the features on the ground.

Also examine on the ground any other features that you have seen on the photographs whose geological cause was not obvious. Their geological significance may now become apparent. Often, photograhs will point you towards places on the ground you might otherwise not have bothered to visit. Some indications on photographs, however, you may never be able to confirm. This does not mean that they do not exist: show

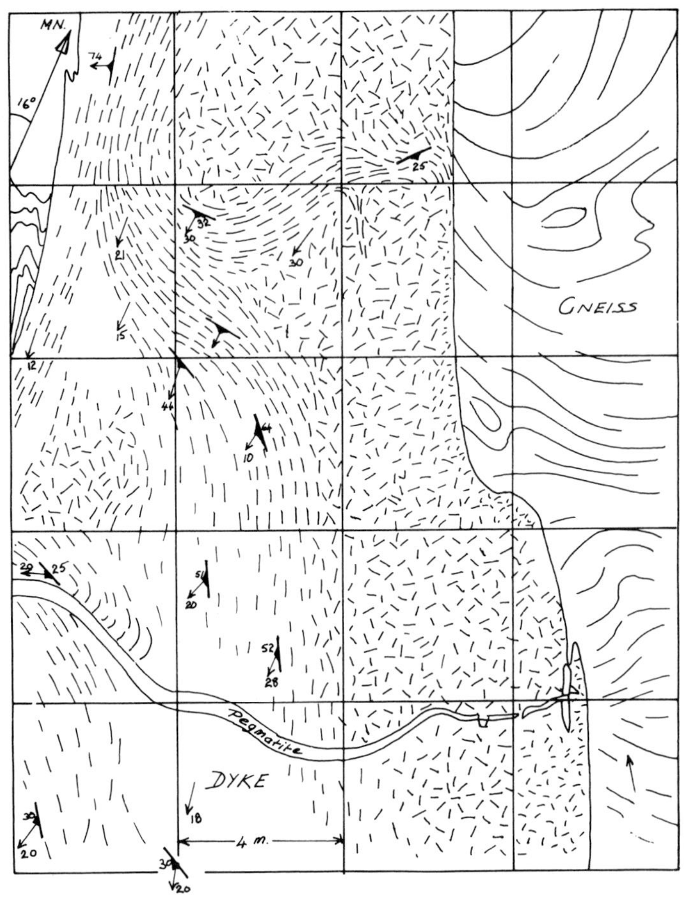

Fig. 4.9 Structures in a deformed 'Scourian' dyke at Badcoll in NW Scotland mapped using the 4 m mesh net shown in Fig. 4.8 (courtesy of R.H. Graham).

them on your field and fair copy maps in purple so that future workers are aware of them. Eventually, their significance may be found. Remember that photogeological evidence is not inferior to other geological evidence, *it is merely different.*

4.9.2 Photogeological features

Only a few indications of what may be inferred from photograhs can be given here. Refer to Ray (1960), Allum (1966), and Lillesand and Kiefer (1979) for further information; but experience is the best teacher.

Tone results from ground reflectivity. It varies with changing light conditions. Sudden changes in tone on a single photograph may indicate a change of rock type owing to a change in vegetation or weathering characteristics.

Texture is a coarser feature caused by erosional characteristics. Limestones have a rough texture: soft shales are often recognizable by a 'micro-drainage' pattern.

Lineaments are any straight, acurate, or regularly sinuous linear features of geologically uncertain significance seen on photographs. They may show in the drainage, as vegetational changes, as ridges, or even as thin lines of lusher vegetation in arid bushland. They may result from faults, master joints, contacts, or for other geological reasons. The cause of some lineaments is never discovered.

Vegetation is an excellent guide to geology and changes can usually be seen more easily on photographs than on the ground. It contributes to both tone and texture.

Alluvium, swamps, marshes, etc. are quite distinctive on photographs and their boundaries can usually be mapped better from photographs than on the ground.

Strikes and dips can be seen from dip slopes, scarp edges and from the way in which beds 'vee' in valleys. There are even methods of calculating the amount of dip where large dip slopes are exposed.

4.9.3 Systematic analysis

Only a brief description of systematic photogeological analysis can be given here.

1. Tape an overlay of *Permatrace* or *Mylar* to one photograph of a stereopair: mark the *pp* and *cj's* on it (Section 3.7.1). (Fig. 4.10a).
2. Under the stereoscope, trace the drainage onto the overlay (in black) to provide a topographic framework. Include alluvium and terrace boundaries. Outline areas of scree, landslide, outwash, etc
3. Trace (in purple) scarp edges and indicate the direction of dip by arrows pointing down the dip slopes—the steeper the slope, the more the barbs (Fig 4.10b).
4. Draw (in purple) known marker beds or beds which can be easily traced. Indicate their dip by 'ticks'—the steeper the dip, the more the ticks (Fig. 4.10b).
5. Show obvious faults in red.
6. Plot as lineaments all linear and arcuate features whose cause is uncertain. Show as purple lines broken at invervals by three dots.
7. Draw contacts in purple as *dotted* lines.
8. Identify rocks and label formations.

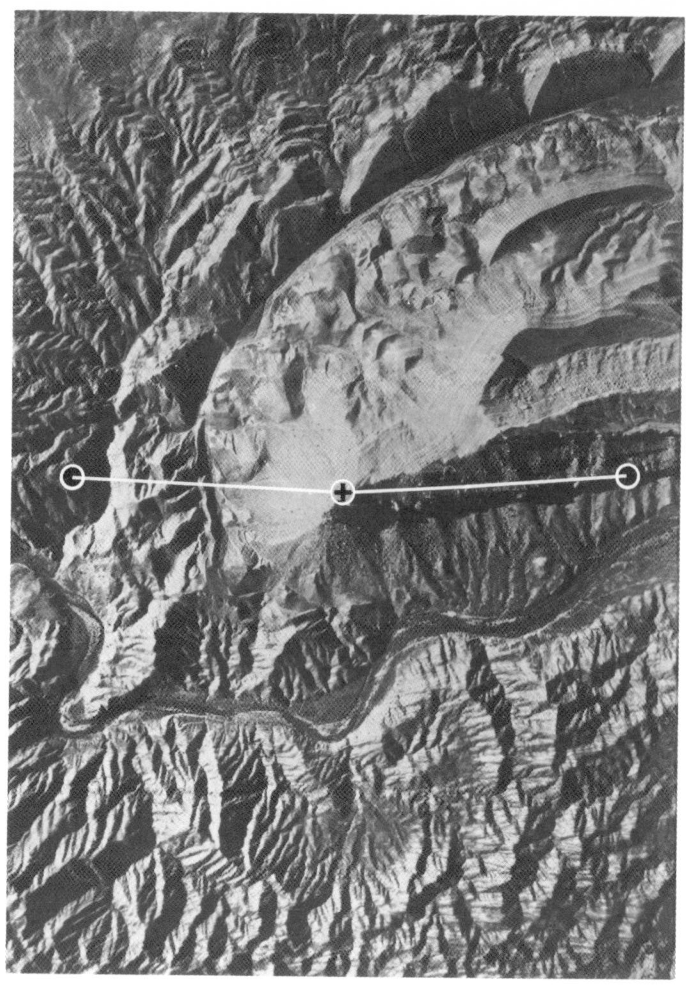

Fig. 4.10 (a and b) Comparison of an aerial photograph (courtesy of British Petroleum Ltd) with its photogeological interpretation.

Scree and alluvium

Middle Fars marls and limestones

Lower Fars limestones

Lower Fars marls with gypsum, anhydrite and salt

Check your interpretation on the ground and against your field map. Amend as necessary and transfer your information to the field map in the appropriate colours to distinguish photogeological from other information. If you are mapping directly onto photographs in lieu of a field map, show any information mapped or confirmed on the ground in black. Always distinguish the two sources of information.

Figures 4.10a and 4.10b show an aerial photograph from Iran compared with its photogeological interpretation. Note that the symbols used are different from those on ordinary geological maps.

5 *Field measurements and techniques*

One object of geological mapping is to elucidate the structural history of the region studied. This can only be done if measurements are made of the attitude of planar structures such as bedding and foliation, and of linear features, such as the trends of minor folds. It is assumed that the reader already knows what these structures are, but many geologists do not know the best way of measuring them. Measurements, once made, must be plotted and recorded, and there are several ways of doing this too, some easier than others. Other structures must also be investigated, specimens collected, photographs taken and, possibly, even soils panned to determine heavy mineral suites where no rocks are exposed. These are all part of mapping technique.

5.1 Measuring strike and dip

Measurements of strike and dip of bedding, cleavage, foliation and jointing are fundamental. Without them, a geological map means little. A useful rule of thumb is to take readings to give an average density of about one for about every 5 cm^2 or 1 $inch^2$ of map surface, regardless of the scale of mapping. Naturally there will be greater concentrations of measurements where strikes vary and fewer where structure is more consistent or exposures poor.

Strikes and dips can be measured in a number of ways, some better than others. Suit your method to the type of exposure. Limestone for instance, often has uneven bedding planes and a method which allows you to measure strike and dip over a wide area of dip surface will give more representative values than one where only a point on the plane is measured. Metamorphic rocks offer additional problems. Measurements of cleavage often have to be made on very small surfaces, sometimes overhanging ones. There may even be more than one cleavage or foliation and at least one of them may be obscure and difficult to measure. You must use your ingenuity. Many gneisses crop out as pavements or turtle-backs where the trace of foliation is clear enough but the dip is difficult to see. Like limestone bedding planes, joints tend to have uneven surfaces; take this into consideration when measuring them. One point must be emphasized: you must plot measurements on to your map immediately after you have taken them, so that any mistakes made in reading your compass—and they do happen—are obvious. Only in very bad weather is it permissible to log readings in your notebook and plot them back in camp. Joints are an exception. They tend to clutter your map without adding to a direct understanding of structure. Record joint directions in your notebook and plot them on to map overlays later, or treat them statistically. An exception to the rule of immediate plotting of structural measurements is where structures are locally complex: then you may draw an enlarged sketch map in your notebook and log the measurements with it. Several different methods of measuring strike and dip are described below. Modify them as occasion demands. Apparent dips can be used to calculate true dip when true dip cannot be measured directly (Sect. 9.2.2).

5.1.1 Method 1

This, the *contact method*, is commonest of all. Use it where the surface to be measured is smooth and even. If there are *small* irregularities, lay your map case on the rock and make your measurements on that. Sometimes, such a small area of bedding or cleavage is exposed that this is the only method which can be used. Place the edge of your compass on the surface, hold it horizontally, align it parallel to strike and read the bearing (Fig. 5.1). Some compasses are provided with a level bubble, so there is no difficulty in establishing strike. With others, you may first have to determine strike with your clinometer, as follows: rotate the clinometer on the rock until it reads zero dip and, if necessary, scratch a line parallel to it with your hammer or lay down your scale. With practice you can usually estimate strike with sufficient accuracy, but where surfaces are nearly horizontal, strike may be more difficult to determine. Then it may be easier to determine the direction of maximum dip and scratch a strike line at right angles to it. Alternatively, if you have water to spare, let a little run over the surface to determine the dip direction. Measure dip with your clinometer at right angles to strike (Fig. 5.2).

Fig. 5.1 Measuring strike by contact (Method 1) using a map case to smooth the surface.

5.1.2 Method 2

On large uneven planes of relatively low dip, estimate a strike line a metre or more long (if necessary mark it with a couple of pebbles), then stand over it with your compass opened out and held parallel to it at waist height. In stream sections or on lake shores nature may help, for a water line makes an excellent strike line to measure. The same method can be used for measuring the strike of foliation or of veinlets exposed on flat surfaces

(Fig. 5.3). Because you measure a greater strike length, it gives more accurate readings than the contact method, and it is particularly useful where foliation is indistinct and seen better in the rock as a whole. Dip is often difficult to measure in some exposures, because there may be no dip surface exposed. The *end-on* method must then be used; sometimes you may even have to lie down to do it. Hold your clinometer at arm's length in front of you and align it with the trace of foliation seen in the end of the exposure, ensuring that the sight line is horizontal and *in the strike of the plane measured.* Figure 5.4 shows a typical exposure suitable for 'end-on' dip measurements.

5.1.3 Method 3

This gives reliable measurements of strike and dip in regions where large areas of moderately dipping bedding planes are exposed or where surfaces are too uneven to measure in any other way. Extreme examples are the limestone dip-slopes often seen in semi-arid countries, but the method can also be used on smaller uneven surfaces, including joint planes. Stand at the end of the exposure (kneel or lie, if necessary) and ensure that your eye is in the plane of the surface to be measured. Sight a horizontal (strike) line across the surface with a hand-level, then sight your compass along the same line and measure its bearing. This will give a reading which averages out the unevenness of the plane (Fig. 5.5). To measure dip, move back so that you can see as much dip surface as possible, then take an 'end-on' reading (Fig. 5.6). Compasses with built-in hand-levels, such as the Brunton, are ideal to establish the strike line for this type of measurement.

5.2 Plotting strike and dip

Plot strike and dip immediately you have measured them. The quickest way to plot a bearing is by P.O.P. (pencil-on-point method) devised by Edgar H. Bailey of the US Geological Survey. It takes only a few seconds, as follows:

1. Place your pencil on the point on the map where the observation was made (Fig. 5.7a).

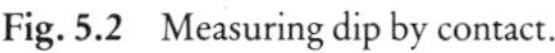

Fig. 5.2 Measuring dip by contact.

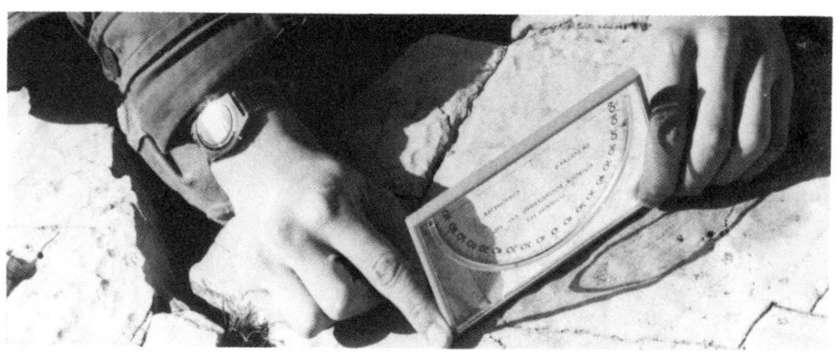

Fig. 5.3 Method 2 used to measure strike of veinlet on horizontal surface.

2. Use your pencil as a fulcrum and slide your protractor along it until the origin of the protractor lies on the nearest north–south grid line; then, still keeping the protractor origin on the grid, slide and rotate the protractor around your pencil still further, until it reads the correct bearing (Fig. 5.7b).
3. Draw the strike line through the observation point along the edge of the protractor (Fig. 5.7c).

The larger the protractor, the better: 15 cm diameter is recommended. If necessary draw extra grid lines if they are spaced too far apart on your field map. Some bearings, such as those lying between 330° and 30°, are easier to plot from the east–west grid lines

The Silva compass has the advantage that, if used according to the directions enclosed with it, you can use the compass itself as a protractor. Make your reading, then without disturbing the setting of the rotating graduated ring, align the N-arrow in-

Fig. 5.4 An ideal exposure for *end-on* measurement of dip.

scribed on the transparent base of the compass case with a grid line and slide it into position (Fig. 5.8).

5.3 Recording strike and dip

It is usually unnecessary and time wasting to enter strike and dip readings in your notebook. It takes little extra time, however, to record the bearing of the strike, in addition to the amount of dip, against the symbol on your field map. This is particularly convenient when mapping on aerial photographs when you must later re-

Fig. 5.6 Measurement of dip of uneven surface by Method 3.

Fig. 5.5 Measurement of strike of uneven surface with a prismatic compass (Method 3).

plot your field information on to a base map of a different scale.

5.3.1 Right-hand rule

Strikes and dips must be recorded in a manner where there can be no possible confusion over the direction of dip: the recording of dip 180° in error is a common mistake. Many geologists write the bearing of the strike, followed by a stroke, and then the amount of dip and the quadrant it points to, viz. 223/45NW. The right-hand rule is simpler: always record strike in the direction your right index finger points when your thumb points down the dip (Fig. 5.9). The quadrant letters can now be omitted

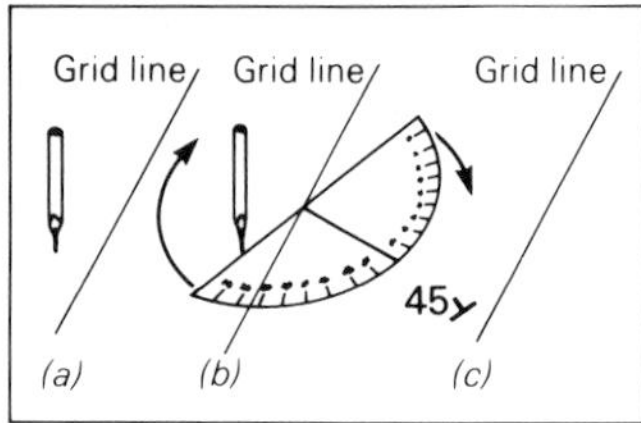

Fig. 5.7 Plotting a bearing by P.O.P. (pencil-on-point).

and the reading of 223/45NW now becomes 043/45. All types of planar information can be written in this form. If you use this method, note the fact in front of your notebook to inform future readers.

5.4 Measuring linear features

Linear features related to tectonic structures are termed *lineations* and the methods of measuring them described here can be used for any other linear feature, whether resulting from glaciation, currents associated with sedimentation, or flowage in igneous intrusions.

5.4.1 Trend, plunge and pitch (or rake)

A lineation is defined in space by its *trend*—the bearing of an imaginary vertical plane passing through it—and by its inclination of *plunge* in that plane (Fig. 5.10). Some lineations appear as lines on an inclined surface, such as where the trace of bedding can be seen on a cleavage plane. These lineations can often be measured more easily by their *pitch* (rake), that is, the angle the lineation makes with the strike of the surface on which it occurs (Fig. 5.11a). Providing strike and dip of the surface has been measured, trend and plunge can then be calculated on a stereographic net. Always log the angle of pitch in your notebook by its clockwise angle so that there is no ambiguity over its direction on the surface (Fig. 5.11b). Pitch can be measured with a common transparent protractor, the bigger the better, a Dr Dollar clinometer or a Silva-type compass.

5.4.2 Measuring lineations

Although some lineations can be measured by their pitch, most must be measured directly with a compass. Sometimes this is simple, as in the case of the stretched conglomerate pebbles shown in Fig. 5.12. All that needs to be done is to stand above the exposure and measure the trend, vertically below. Plunge is then meas-

Fig. 5.8 Plotting a bearing with a *Silva* compass.

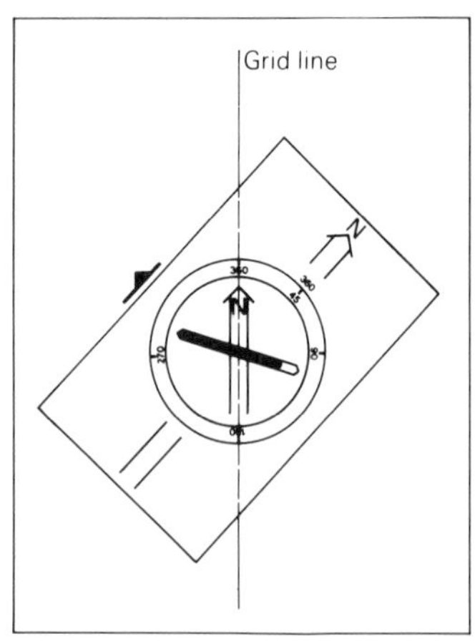

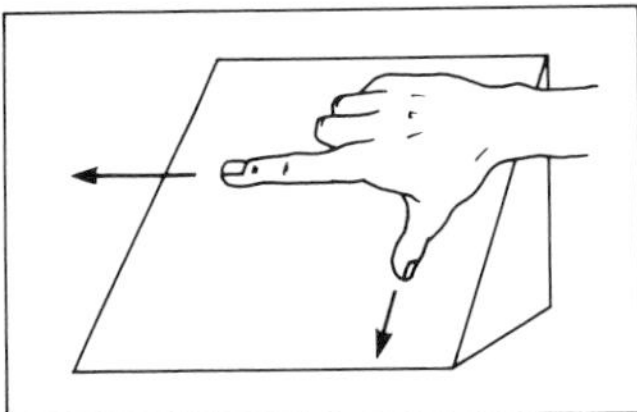

Fig. 5.9 The *right-hand rule* for recording strike and dip.

ured by 'contact' or 'end-on' methods. Direct measurement of trend and plunge can also be made for lineations on gently to moderately dipping surfaces but, as surfaces become steeper, it is increasingly difficult to measure trend accurately. Figure 5.13 shows one way it can be done if your compass is suitable. Lay the edge of the compass lid along the lineation; level the compass case by noting whether the compass card or needle floats horizontally (some instruments have a circular level bubble. If the compass case is truly horizontal the edge of the compass must, geometrically, lie in the trend plane. Read the bearing for trend. Plunge is measured by direct contact in the trend plane. Very serious errors in trend may arise from measurements merely 'eyed-in' from above. Lineations can be measured accurately and easily by the Japanese compass illustrated in Figs. 2.3e and 2.6.

Some lineations are most difficult to measure, especially those related to minor folds. Considerable ingenuity may be needed. Folds in gneisses may often, at first sight, appear to show up beautifully, but on closer examination, it may be found that no crest or hinge lines are properly exposed (Fig. 5.14). It is these crest or hinge lines that you normally measure. Figure 5.15 indicates some of the considerations which must be kept in mind.If the axial plane of a fold is vertical, then the crest and hinge lines are coincident and the trace of the axial plane indicates its strike, whether the fold plunges or not (Fig. 5.15a). If, however, the fold is overturned, the axial plane is no longer vertical and the hinge line now becomes a lineation formed by the intersection of two surfaces—the inclined axial plane, and the vertical plane in which the trend of the hinge is measured (Fig. 5.15b). In practice, what you can usually measure are: the trace of the axial plane on the rock surface, the trend and plunge of the hinge, or the trend and plunge of the crest line. Only too often, the plunge itself cannot be measured at all. You may be able to estimate it, but sometimes you can do little more than indicate its direction, and whether it is gentle, moderate or steep.

Fig. 5.10 Geometry of trend and plunge.

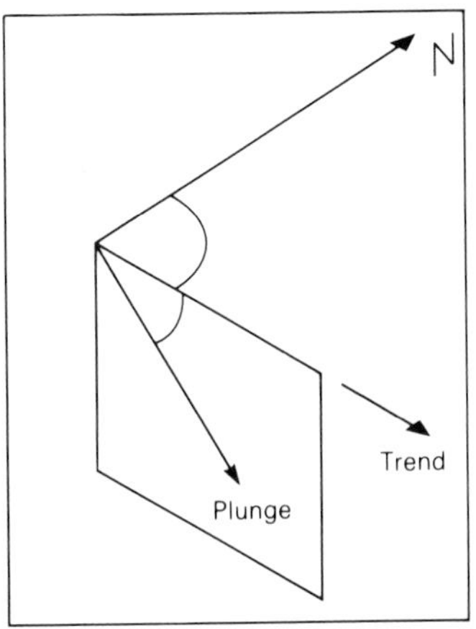

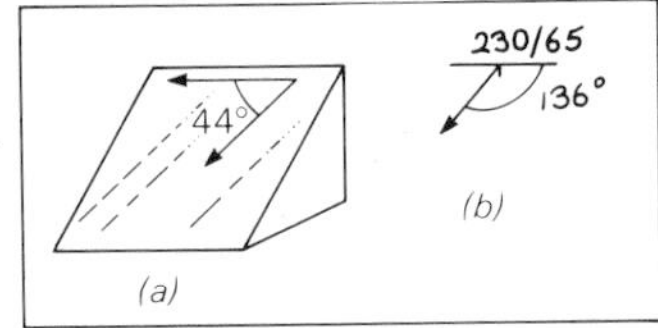

Fig. 5.11 (a) Geometry of pitch on a lineated dip slope. (b) Record pitch in your notebook by a diagram: record also strike and dip.

5.5 Folds

Minor folds are quite frequently seen in outcrop, major folds seldom are except in the more arid countries. Minor folds can, however, often provide the key to the major folds they are related to. They reflect the shape and style, and their cleavage indicates the attitude of the axial planes, of the major folds. Also, their vergencies indicate where the closures of the major folds lie and the attitudes of their axes and axial planes: for example, the 'Z' fold shown in Figure 5.16 indicates that the major antiformal fold closure is to the right of the picture, the synformal to the left. It also indicates overturning of the axial plane. Minor folds, such as this, are too small to show in outcrop on your geological map except as a symbol selected from the list of symbols printed inside the front flap.

The terminology of folds is complex; it is also often ambiguous and before going into the field you are well advised to read Fleuty's paper, *The description of folds* (1964). In general, map the directions and inclination of axial planes of folds where it is possible to do so, and note fold shapes, attitudes and sizes. Measure any cleavages related to them and all lineations and intersections of cleav-

Fig. 5.12 Stretched conglomerate pebbles in East Africa: trend and plunge can be measured directly.

Fig. 5.13 Measuring lineation on a steep surface with a hinged-lid compass (see text).

ages, such as those with bedding. Show by symbols the trends, plunges and shapes of all folds too small to show in any other way. Make sketches.

Fleuty (ibid.) gives numerical values for terms defining the attitudes of folds, etc., open, close, tight, etc. In the field, *make measurements* wherever you can, and avoid terms such as gently, moderately, steeply plunging, in your notebook. Make sure that you are well prepared in the basic concepts of structural geology and keep a textbook on the subject in camp with you. Much of the difficulty you will encounter is in recognizing structures in the field when you see them for the first time: they seldom resemble those idealized diagrams in textbooks.

Fig. 5.14 Minor folds in Precambrian granite gneiss in East Africa.

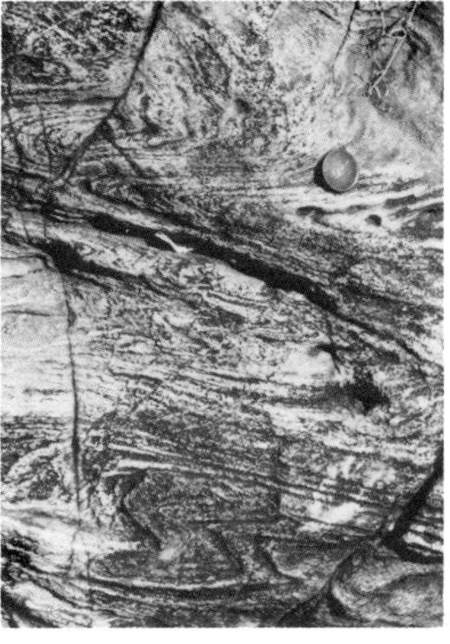

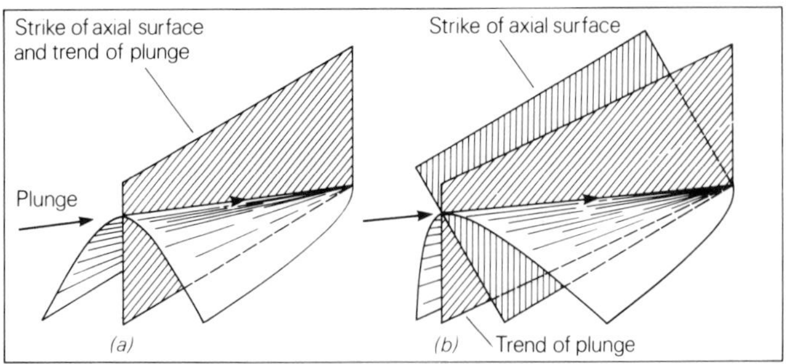

Fig. 5.15 (a) Axial surface in an upright plunging fold. (b) Overturned plunging fold: note that now the *trend* of the hinge-line is no longer parallel to the *strike* of the axial surface.

5.6 Faults

Most faults are never mapped because they are never seen. Many have such small displacements that it matters little if they are individually missed, but record those you do see to help you to establish a fracture pattern. Major faults are more likely to be found, but even those with displacements of tens of metres may be missed where exposures are poor. Many faults have to be mapped by inference. Suspect a fault where there are unaccountable changes in lithology, where sequences are repeated, where strikes of specific beds cannot be projected to the next exposure, or where joint spacing decreases suddenly to a few centimetres, and rocks become flaggy. Topography is often a good guide. Faults may result in boggy hollows, seepages, or, in semi-arid countries, even a line of taller, greener trees surrounded by lower flat-topped acacia. Most fault zones erode a little faster than the adjacent rocks to form longitudinal depressions, but beware, some faults in limestones may form low ridges owing to slight silicification which helps to resist erosion. Faults are most easily traced on aerial photographs where the vertical exaggeration of topography seen under the stereoscope accentuates those minor linear features called *lineaments*— features often difficult to find on the ground.

The sense of displacement of a fault, that is, distinguishing the downthrown side, may become evident only by noting the difference in stratigraphy or lithology on each side of a fault. In textbooks much is made of slickensides and, if they are seen, they should be measured. But do not put too much faith in them—they merely reflect the last phase of movement, and most faults have several phases, not always all in the same direction. Note also that faults have a thickness which may be wide enough to plot on large-scale maps, particularly on sketch maps. They may also be gouge-filled or breccia-filled, or they may be mineralized, and their outcrops are seldom as straight as shown on maps. Record all such observations in your notebook.

5.7 Thrusts and unconformities

Thrusts and unconformities are treated together because one can easily be mistaken for the other. Large thrusts are often obvious, with older rocks overlying younger; but not all thrusts show such a clear relationship. Sometimes thrusting may be discovered only by unexpected changes in stratigraphy. If the thrust surface is not properly exposed, the upper and lower 'plates' may both show angular unconformity with its assumed position, or the surface may show complete disregard for the stratigraphy of the upper plate. If the surface is exposed, the situation should be clearer. The lower part of the upper plate should not show any of the sedimentary features expected in a stratigraphic unconformity; there may be shearing along the surface, or mylonites. Where mylonite does occur it may be thick enough to map as a formation in itself and form a useful marker.

Not all thrusts are major thrusts. Some are merely reverse faults, others may form imbricate zones, consisting of numerous small sub-parallel thrusts associated with larger thrusts, as in the Scottish *Moine*. Such zones are marked by multiple repetitions of partial sequences which, if poorly exposed, are impossible to map completely. Sometimes the spacing between individual thrusts may be only a few metres, sometimes tens of metres.

Stratigraphic unconformities show younger rocks lying on older rocks below, usually, with angular unconformity between them. The rocks just above an unconformity should also

Fig. 5.16 Minor fold in Precambrian sedimentary rocks in East Africa. The major antiform closes to the right.

show features which indicate original deposition on an eroded surface. Unfortunately, this relationship is not always as clear as textbooks suggest, especially where rocks have been metamorphosed. Sometimes, to confuse matters, there is angular unconformity on both sides of the break if the later rocks have been deposited on a sloping surface (Fig. 5.17). A *disconformity* may be even more difficult to recognize. It represents a break in sedimentation and the beds are parallel both above and below it. It should be discovered during sedimentary logging by the evidence of erosion between the two stages of deposition.

5.8 Joints

Joints occur in every type of rock, sedimentary, pyroclastic, plutonic, hypabyssal, volcanic and metamorphic. Record joints, but do not clutter your map with them. Enter them into your notebook and replot them on to transparent overlays to your fair-copy map, or plot them as statistical diagrams, such as stereograms and rose diagrams in equal area 'cells', spread over the surface of your map overlay. Master joints, those dominant major joints, are an exception. They may sometimes warrant being shown on your map. Follow them on the ground or on aerial photographs and plot them in a similar manner to faults, but with the appropriate 'joint-dip' symbol. In general, however, keep joints off your maps.

Measure the strike and dip of joints in the same way as bedding. Often

Fig. 5.17 Unconformable Neogene conglomerates lying on Palaeozoic limestone and phyllite in Turkey. Left-hand arrow indicates limestone/phyllite contact. Right-hand arrow indicates conglomerate/limestone unconformity.

surfaces are uneven and contact methods unsuitable. Book readings in your notebook, using the *right-hand rule*, together with estimates of their lengths. Note also the spacing between joints in each set and what formations each set penetrates. Master joints may show up well on aerial photographs, especially in limestone regions where they may be indicated by karst patterns and by lines of sink-holes. Joint patterns seen on photographs can sometimes be used to distinguish one formation from another.

5.9 Specimen collecting

Collect representative specimens of every formation and rock type you show on your map. Often, several specimens of the same formation are needed if it varies in composition over the region. Even if it does not vary, you may need specimens from different parts of the region to prove that it does not. Some variations in composition may not, of course, be obvious in hand specimen and so extra specimens are needed as a safeguard. The size of specimen you collect must depend on the purpose you wish to put it to, not on what you think you can carry. See your rock cutter *before* you go into the field to see what he needs for thin-sectioning. Whenever possible, collect material which shows both weathered and fresh surfaces and, if necessary, take two samples to show both. Do not collect just any piece of rock you can knock off an exposure with your hammer. The easiest piece to break off may not represent the exposure as a whole. You may have to spend considerable time in breaking out a good specimen, with hammer and chisel.

Having broken off a specimen, trim it. Mark sedimentary rock specimens to show which is their top. Metamorphic specimens may need to be oriented so that directional thin sections can be cut: break them off, and then fit them back to where they were broken off and mark a direction and 'dip' on the surface so that they can be oriented in space when thin sections are cut from them. Whenever possible orient specimens before breaking them off.

5.9.1 *Marking specimens*

Rock specimens are best marked with a waterproof felt-tipped pen or, for dark rocks, a yellow timber crayon or a numbered piece of surgical 'sticking plaster'. Wrap specimens in newspaper to protect them from bruising and to spare your rucksack. In camp, scrub your specimens, dry them and then add a spot of white paint: later number the spot with black paint. Re-wrap your specimens in newspaper and number the packets on the outside with a felt-tipped pen so that they can be easily identified should you want to examine any of them again in camp.

5.9.2 *Fossils*

Some fossils are easy to remove from their parent rock, others are not. Many are embedded with only a small portion showing: scrape away enough rock with a knife to see whether the specimen is worth collecting, and then cut out the piece of rock containing it. Many other fossils are casts or impressions in the rock: again collect the piece of rock containing them. Wherever possible collect both external and internal casts: both are important. Sometimes you may have to collect several kilograms of fossiliferous rock so that the individual fossils can be extracted in the laboratory. This is particularly so where micro-fossils are needed. Mark all specimens with the way up in which they are found and do not collect more material than you need.

Pack delicate specimens in boxes or tins and pad them with cotton wool, tissue or newspaper, or use expanded polystyrene ceiling tiles cut to fit the boxes. Carry a selection of containers, from matchboxes upwards in size. Wrap non-fragile specimens in newspaper and treat in the same way as rock specimens.

5.9.3 *Booking specimens*

Log specimens in your notebook immediatley you have collected them. Preferably write their number in the left-hand margin, so that their details can be easily relocated. If specimen numbers are written in red pencil, they can be distinguished from observation numbers listed in the same column. Alternatively, if you are collecting large numbers of specimens, add a column to your notebook specially for this purpose. In addition to logging specimens on the working pages of your notebook, register them in the back of the book too. This avoids finding yourself with two almost identical specimens with the same number and no way of telling which is which. A register also helps you ensure that you have collected specimens of everything you should collect, and if you show the page numbers where they are more fully described in your notebook, it acts as a handy ready-reference (Fig. 5.18).

5.9.4 *Shipping specimens*

Geological specimens are heavy and i shipped in a box which is too larg can only be accepted as freight. Smaller boxes, which one man can lift, car go much more quickly by passenge transport. A box about 25 × 30 × 2 cm made of +1 cm timber, battened and banded with steel tape, is acceptable by T.I.R., railway passenger ser vices, and airlines. Mark your nam and address on the top and sides, an add ROCK SPECIMENS FO SCIENTIFIC RESEARCH. Neve write 'ore specimens' or 'minera specimens' on boxes. Most countrie do not appear to have export regula tions controlling *rocks*, but do fo *minerals* and *ores*. 'Rocks', an hones declaration for any geological mate rial, avoids bureaucratic delays an gets your rocks back to your labora tory more quickly.

5.10 Field photography

A camera is an asset in the field and 35 mm model is the most versatile The only important specification i that it should not have a fixed focu lens. The selection of lenses depend on your pocket. A 50 mm lens is prob ably the most useful; a wide-angl lens is probably the next choice if yo can afford more than one. Always us an ultraviolet filter when using colou on a shore or in mountains ove 2000 m. It is virtually colourless an can be kept on the camera at all time so that it is not forgotten. Whe photographing exposures in blac and white, a dark yellow filter help to bring out detail. 'Instant' camera have the advantage that an exposur may be photographed and annotate with a fibre-tipped pen on the spot.

(89)

SPECIMEN REGISTER

Spec. Nº		Page
A 1	Part ox. ore – Alamkandi	14 a
A 2	Grey laminated bedded Lms	
A 3	from benches at IV Δ. lams	15a
A 4	10-20 cm thick	
A 5	Gossan from Δ IVH Hillside	16
A 6	Br. fossilif Lms from Δ IVH hilltop	16
A 7	Grey lam Lms from Δ IVH hilltop	16
A 8	—— " ——	16
A 9	Massive un-lam grey lms from Δ IVD	16a
A 10	Smithsonite (hydrozincite?) from Dump	16a.
A 11	—— " ——	16a
A 12	Lo-grade ore from pit	16a
A 13	Hi-grade ore from pit	16a.
A 14	Brecc – red rock fragst Zn CO_3	17.
A 15	Brecc – ore in phyllite	17.
A 16	Ore from dump – hi-grade?	17.
A 17	Grab-samples from dump	17.
A 18	Calc-chlorite schist.	17A
A 19	Chlorite sch	17A
A 20	Sericite sch	17A
A 21	Cobaltite/erythrite – Memshem.	19
A 22	Cobaltite	19
A 23	Barite?	19
A 24	Malachite stained carbs	19
A 25	Lam Lms showing weathered surface	19
A 26	Gossan from Δ IV hillside	19a
A 27	Amphib float —— " ——	19a
A 28	Ser-phyll from Δ I h'side	19a
A 29	Chloritoid (?) sch —— " ——	20a
A 30	Serp-talc schist	20a

Fig. 5.18 Specimen register in a field notebook.

Whenever you take a picture, make a sketch of the scene in your notebook to show what to look for in the photograhic print. This is particularly important for photographs of rock exposures for you may not see the prints until you have returned from the field. It may then be difficult to identify photographs, particularly where you have taken several of similar rock exposures. Log every picture taken and number it in the same way as you record specimens, entering the number in blue pencil in the left-hand column of your notebook, or in the 'specimen' column. Keep a photograph register, as for specimens. To keep track of photographs of exposures, make a device from two strips of perspex taped together between which you can slip large numbers cut from a trade calendar, as in Fig. 5.12. Whatever method you use to photograph outcrops, you must include some object to give it a scale.

Photographs can be taken in black and white, or in colour, either as prints or transparencies. The choice depends on what you intend to use your photographs for. Colour prints are effective for scenery but can be disappointing for close-ups of outcrops. For general geological work, black and white photography is probably adequate, especially if you intend to publish the photographs in reports or papers.

There is no need to mark on your map where you have taken photographs of exposures: these will already be logged against an observation number in your notebook. It is, however, helpful to indicate the direction in which you took a scenic photograph by an arrow on your field map, so that you can identify topographic points more easily later. Finally, having taken photographs in the field, file your negatives so that they can be easily found again.

5.11 Panning

Every geologist should be able to use a gold pan. It needs little practice. Gold and cassiterite can be 'panned' for, but many rock minerals which survive erosion can be concentrated by panning too. These include garnet, rutile, zircon, epidote, monazite, magnetite, haematite and ilme-

Fig. 5.19 Panning in the River Euphrates

nite. Differences in the 'heavy mineral suites' extracted by panning soils are useful guides to the underlying geology in poorly exposed regions (see *loaming*, Section 4.4.4).

Unlike gold, garnet and epidote, etc., are only a little more dense than the sand and rock debris accompanying them (S.G. 3.2–4.3 compared with ±2.7) and more skill is required to concentrate them. A 30 cm diameter pan is sufficient for purely geological purposes. Keep it spotless and free from rust and grease. Collect *stream gravel* from the coarsest material you can find, for that is where the heavier minerals concentrate. Dig for it with a trowel or entrenching tool and get down to bedrock if possible. Collect *soils* from below the humus. Heap the pan full of material, then shake vigorously under water, in a stream, or even in a tin bath. The finer heavies will pass down through the lighter coarse material, a process known to the mineral dresser as *jigging*. Larger pebbles can be scraped off the top and thrown away. Continue shaking and removing pebbles until only sand size material is left. Then, tilt the pan away from you, dip it into the water and swirl the water around the pan so that it just washes off the lighter sands. Give an occasional shake to ensure that the heavier minerals can work their way down into the angle between the bottom and the side of the pan. When little is left except dark and coloured minerals, alternate shaking from side to side with letting water flow over the concentrate to wash away any remaining lighter sands (Fig. 5.19). Finally, let in a little water and give a single swirl around the pan sides to produce a 'tail' of minerals graded in order of their density. Examine the tail with a handlens under a shallow cover of water. Identify any minerals you can, then collect the concentrate in a phial for future examination. Panning is like fishing: you do not actually have to catch anything to enjoy it!

6
Rocks, fossils and ores

This chapter assumes that readers are already familiar with systematic methods of rock naming in the laboratory, know basic palaeontology and can recognize the materials mentioned. Here, you are told how to apply that knowledge in the field.

6.1 Rock descriptions

When you have mapped a rock unit for long enough to be familiar with it, describe it fully and systematically in your notebook. Rock descriptions are essential when you come to write your report. A rock description made from memory is unlikely to be accurate or complete. One made in the field describes the rock as seen, with measurements of specific features and factual comments on those subtle characteristics that are impossible to remember properly later. It also ensures that you record *all* the details needed.

Systematically describe each rock unit shown on your map in turn. Preferably, work from the general to the particular. Describe first the appearance of the ground covering it, its topography, vegetation, land use, and any economic activity associated with it. If the soils are distinctive, describe them too. Next describe the rock exposures themselves, their size, frequency and shape; whether they are turtle-backs, pavements, tors or ridges, jagged or rounded. Comment on joint spacing, bedding and lamination (see Appendix III.), structures and textures, cleavage and foliation. Support your observations with measurements. Describe the colour of the rock on both weathered and freshly broken surfaces. Weathering often emphasizes textures; note its effect, such as the honeycomb of quartz left on the surface of some granites after feldspars have been leached away, which immediately distinguishes silicic from less silicic varieties. Finally, describe features seen in hand-specimen, both with and without a handlens. Note texture, grain-size and the relationship between grains. Identify the minerals and estimate their relative quantities. Name the rock. Where appropriate, prepare a sedimentary section and/or log (Sections 6.3.2, 6.3.3). A *formation letter(s)* will eventually be assigned to every mappable rock unit, but that is something to be done later. Remember you can take a specimen home with you, but not an outcrop. Ensure that you have all the information you need before you leave the field.

6.2 Identifying and naming rocks in the field

There are two problems here. The first is to find out what the rock is in petrographic terms, the second to give it an identifying name to use on your fair copy map and in your report. The first is the *field name*, the second, the *formation name*.

6.2.1 Field name

A field name should be descriptive. It should say succinctly what the rock is—but you cannot name a rock until you have identified it. A field geologist should be able to determine the texture, the relationship between minerals, identify the minerals, and estimate their relative abundances in most rocks, under a handlens. He should be able to distinguish orthoclase from plagioclase, and augite from hornblende, in all but the finer-grained rocks. He should be able to give some sort of field name to any rock. Dietrich and Skinner's *Rocks and Rock Minerals* (1979) is an excellent guide to identifying rocks in this manner.

A field name should indicate structure, texture, grains-size, colour, mineral content and the general classification the rock falls into, e.g., *thin-bedded fine-grained buff sandstone* and *porphyritic medium-grained red muscovite granite*. These are full field names and shortened versions or even initials can be used on your field map. Avoid at all costs calling your rocks *A, B, C*, etc., on the assumption that you can name them properly in your laboratory later. This is the coward's way out. If you are really stuck for a name, and with finer-grained rocks it does happen, then call it *spotted green rock*, or even *red-spotted green rock* to distinguish it from *white-spotted green rock*, if need be. Ensure, however, you have a type specimen of every rock named. Sometimes, you may even find it helps to carry small chips around with you in the field, for comparison.

6.2.2 Formation names

Consider the *formation* as your basic mappable unit—the one you show with a specific colour or pattern on your map. Some formations may consist of a number of different *members*, each of which has to be identified and given a field name although they cannot be mapped in separate colours. The formation must be given an identifying name. This may be purely descriptive of its geological position, such as the *Boundary Quartzite* which forms the base of a system in Africa, or it may have a locality name, such as the *Igara Schist*, which lies unconformably below it, or the *Chitwe Granite*, which intrudes both. Both the Igara Schists and the Chitwe Granite contain several different rock types, but each is distinguished on the map by only one colour. Formation names may have to be modified later, as further work is done at larger scales. Some formations, usually igneous rocks, can be distinguished only by a rock name, such as *dolerite, elvan, quartz-diorite*. Others may be named after a locality or the group they commonly intrude, such as the *Karroo Dolerites*.

6.3 Naming and describing sedimentary rocks

Formal rules are laid down for naming sedimentary sequences. Any description of a succession should be accompanied by a *stratigraphic sec-*

tion to define the sequence and, in detailed work, a *sedimentary* or *graphic* log to illustrate the variations in sedimentation.

6.3.1 *Sedimentary formations and members*

A sedimentary *formation* has 'internal lithological homogeneity, or distinctive lithological features that constitute a form of unity in comparison with adjacent strata'. It is the basic mappable unit. For convenience, it may be sub-divided into *members*. If a formation has not already been formally named, name it yourself in the approved manner, attaching a place name to the rock name, e.g. *Casterbridge Limestone Formation*, or for working purposes, just call it the *Casterbridge Limestone*. Avoid loose terms, such as the *White Limestone* or *Brachiopod Bed*. Establish a type section for every named formation for reference or comparison in case problems arise. The Geological Society of London has issued a guide on the subject (Geol. Soc., 1972) and the US Geological Survey offers similar advice (Cohee, 1962).

6.3.2 *Stratigraphic sections*

Stratigraphic sections show the sequence of rocks, distinguishing and naming the formations and members that comprise them. They show the thickness of units, the relationships between them, any unconformities and breaks in succession, and the fossils found. It is impossible to find one continuous exposure that will exhibit the complete succession of a region—even in the Grand Canyon—and a complete section is built up from a number of overlapping partial sections. There may even be gaps where formations are incompletely exposed.

Sections can be measured in a number of different ways and some guidelines are given here. The first task is to select a suitable place with good exposure. Make measurements of the true thickness of the beds, starting at the base of the sequence, and log them in your notebook as a vertical column. In measuring thickness, corrections must be made for the dip of the beds and the slope of the surface on which they crop out. This can be done graphically on squared paper, or trigonometrically (Fig. 6.1). Compton (1966) illustrates various methods of measuring true thickness directly.

Indicate on the stratigraphic section the name and extent of every lithological unit, together with the rock types which comprise it. Take specimens of everything logged. Mark and note the position and names of any fossils found. Collect specimens, where necessary, for later identification. Indicate the position of the section on your field map. Redraw the section from your notebook on squared paper in camp. Later the formation may be simplified and

Fig. 6.1 Correcting for the true thickness of a bed. The stratigraphic thickness AB = AC Sin $\alpha + \beta$.

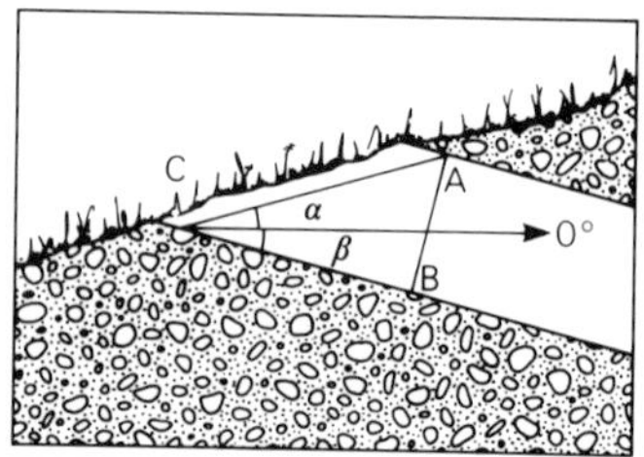

combined with sections from other parts of your mapping area as *columnar sections* or *fence* diagrams (Sections 9.3 and *9.4.4*). Stratigraphic sections may also contain igneous and metamorphic rocks.

6.3.3 Sedimentary (graphic) logs

Although there are similarities, sedimentary logs and stratigraphic sections differ in their purposes. Sedimentary logs are detailed graphic displays of the lithologies, sedimentary structures, and fauna in a succession. The succession is broken down into homogeneous units—termed *sedimentary facies*—which contain distinctive combinations of features. The manner of deposition of a unit can be inferred from its facies, and the overall environment of deposition from its vertical and lateral associations. There are a number of conventions in recording logs. As with stratigraphic sections, the thickness of beds is shown to scale in a vertical column. However, in a sedimentary log there is also a horizontal scale: the width of the column is a measure of the grain-size of each rock unit portrayed (Fig. 6.2). Symbols are used to indicate a wide variety of sedimentary features, such as different forms of ripples, cross-bedding, rootlets and mud-flakes. So far, no convention of symbols has been universally accepted. Devise your own in a form which makes your logs easy to understand.

Choose a site for a sedimentary log as for a stratigraphic section. Measure the thickness of each lithological unit and record its sedimentological features in your notebook. Take special note of the nature of boundaries between units, whether they are erosive, sharp or gradational and see whether there are any lateral variations. Tucker (1982) gives full details.

6.3.4 Way-up of beds

Symbols indicating which way beds 'young' are frequently omitted on maps in strongly folded areas, despite often abundant evidence. There are three main ways of telling which way up a bed is. *Sedimentological* indications are the most abundant and include cross-bedding, ripple marks, sole marks, graded bedding, down-cutting erosive boundaries, load casts, and many others. *Palaeontological* evidence includes trace fossils; burrows and pipes left by boring animals; and roots of crinoids and corals in their growing position. Many palaeontological pointers to way-up are fairly obvious but one on its own is not always reliable. Look at a number of different ones before making a decision.

A most important *structural* guide to way-up is the angle between bedding and cleavage. If bedding dips more steeply than cleavage the fold limb has been overturned (Fig. 6.3). This can be used in rocks devoid of fossils or sedimentological features, even in rocks such as quarzites. In structurally-disturbed areas where it is often difficult to tell which way up beds are, mark the 'overturned' symbol for dip and strike on your map where overturning is confirmed; where the beds are known to be right-way-up, add a dot to the pointer of the usual symbol (see list of geological symbols inside front flap); uncommitted symbols then indicate lack of evidence either way. Wherever there is evidence of way-up in such areas, note what it is, such as *c.b.* for cross-bedding, *r.m.* for ripple

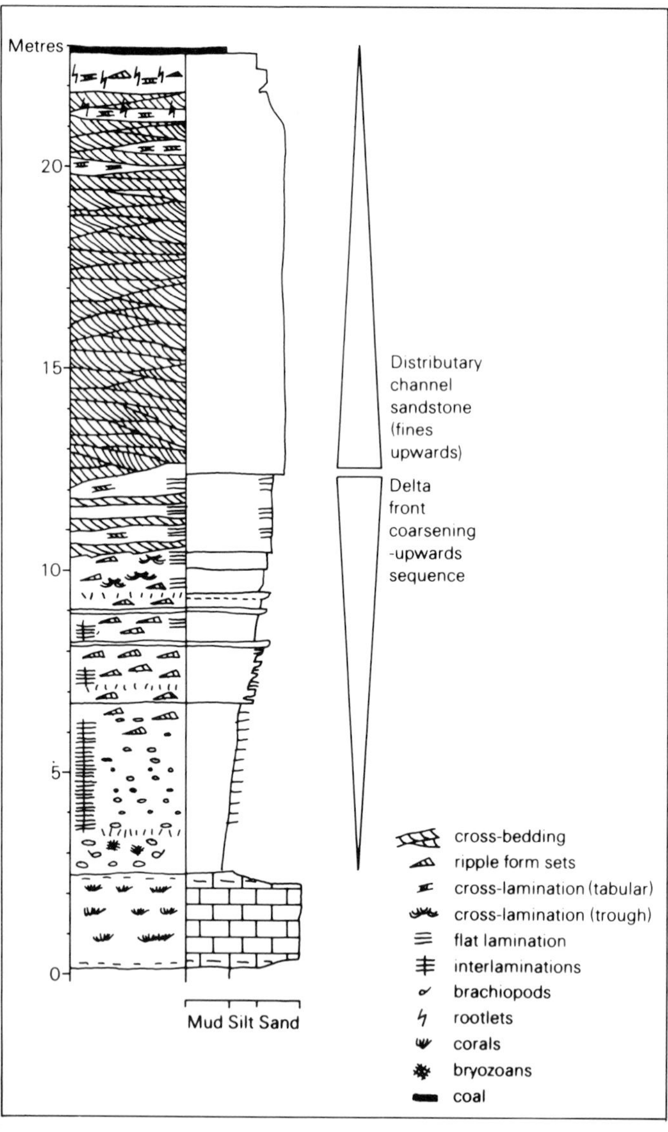
Metres
20
15
10
5
0
Mud Silt Sand
Distributary channel sandstone (fines upwards)
Delta front coarsening -upwards sequence
cross-bedding
ripple form sets
cross-lamination (tabular)
cross-lamination (trough)
flat lamination
interlaminations
brachiopods
rootlets
corals
bryozoans
coal

marks, *t.p.* for trumpet pipes. That is part of your field evidence.

6.3.5 *Grain-sizes*

The vast majority of sedimentary rocks are classified by their grain-size. Anything greater than 2 mm is *gravel*, anything less than 0.0625 mm (1/16 mm) is *mud*, what lies between is *sand* or *silt*. Each of these groups is subdivided into coarse, medium and fine, etc. (Appendix III, Table III.2 inside back cover). Measure larger grains in the field with a transparent plastic scale placed over a freshly broken surface. Use a handlens with the scale for finer sizes. Generally, if a piece of the rock is gritty between your teeth, then silt is present, and if grains lodge between your teeth, there is fine sand, but that should be visible under your handlens.

6.3.6 *Smell*

Some rocks are sandy rocks which contain clay. Breathe on the rock and note whether it returns a clayey smell. This is not infallible, for if the rock is too indurated, the clay minerals will have been altered to new minerals. Other rocks, namely those which once had a high organic content, emit a sulphurous smell when hit with a hammer.

6.3.7 *Hardness*

Always test a fine-grained rock first by scraping your hammer over it. If it scratches, it is probably a sedimentary rock; if it does not, it may be a chert or a hornfels, or an igneous or pyroclasitc rock. Some compact cream, white and grey rocks can be scratched with your finger nail. They are probably gypsum or anhydrite, or possibly rock salt: one lick can settle that!

6.3.8 *Acid*

Every geologist should carry a bottle of 10% hydrochloric acid with him. To use it, break off a fresh piece of rock, blow off any dust, and add *one* drop of acid. If reaction is vigorous, the rock is *limestone*. If it does not fizz, scrape up a small heap of powder on the surface with your knife and add another drop of acid. Gentle reaction indicates *dolomite*. Many carbonate rocks contain both calcite and dolomite, so collect specimens for staining when you return to base. Remember, however, some other, if rarer, carbonates react to acid too.

6.4 Fossils

Fossils cannot be considered in isolation from their environment. All the features found in a fossiliferous rock must be recorded if you are to gain the full benefit from a fossil itself. Note the abundance of the fossils in each fossiliferous horizon of the locality: are they widespread or clustered into groups? Did the fossils die where found, or were they transported there after death? Do they show alignments due to currents? Different fossils may occur in different parts of the same horizon and there may be lateral changes which can be traced over considerable dis-

Fig. 6.2 A graphic sedimentary log. The horizontal scale is a measure of grainsize. Divisions are unequal because the Ø-scale range for silt is only 4/5ths that for sand (see Appendix III). The vertical triangles to the right indicate coarsening and fining in the sequence. (From N. Wales, courtesy of A.R. Gardiner.)

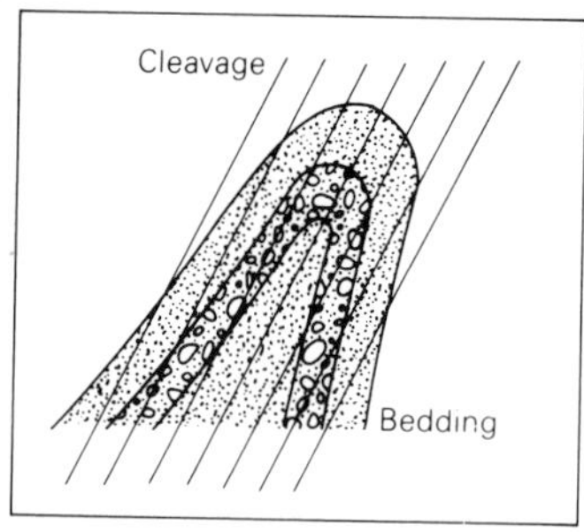

Fig. 6.3 Relationship between cleavage and bedding in an overturned fold.

tances, indicating a changing environment. There may also be vertical changes as the depth of water the rocks were deposited in changed. All this must be recorded in your notebook, either on a measured section or, if the occurrence is suitable, on a stratigraphic section or graphic log.

Do not be over-anxious to collect a fossil when you find it. First study it in place, noting its attitude and surroundings; make notes with sketches if necessary. Probably you will see only a small part of the fossil, perhaps because only a small part of it is exposed, or because only fragments occur. Decide how best to remove it from the rock, then remove the specimen carefully, trying to keep it intact. Use a chisel or even scrape around it with your knife. Sometimes it is better to remove a large piece of rock and carry it around all day, than to be too ambitious in trying to remove a specimen in the field. If you find a whole fossil, one specimen of that species is probably enough; leave the rest for others. Usually, however, you will only be able to collect incomplete fossils. Some may show external features, some internal casts: collect both. As with rocks, name fossils in the field but before going into the field, refer to the types you may expect to see in the rocks you will be looking at. Do not be discouraged if you cannot name in detail every fossil you find. Expert help is often needed.

Once you have discovered a sequence containing specific fossils in one part of your area, you may then find that you can use it for mapping on a wider basis, especially where you have a series of repeated sequences, or cyclothems. The fossils will tell you which part of the series you are in. Again, where you have beds of great thickness, your fossiliferous horizon will tell you whereabouts in that bed you are. It can even tell you the displacement where both sides of a fault are of the same rock type.

Mark good fossil localities on the field map with a symbol, so that you can find them again, but you need not show them all on your fair copy. Mark only the more important localities people may want to visit.

6.5 Phaneritic igneous rocks

Phaneritic igneous rocks are easily recognized and acid to intermediate varieties can usually be readily named. Dark coloured (melanocratic) phanerites are more difficult to identify but you can usually put some field name to them which is nearly correct. Before you go into the field try to look at specimens of the types of rocks you expect to encounter, taken, if possible, from the area you are going to.

6.5.1 Grain-size in phaneritic rocks

Grain-size terminology in igneous rocks differs from that used for sediments, namely:

Coarse grain	+5 mm
Medium grain	1–5 mm
Fine grain	–1 mm

Use the terms coarse, medium and fine when discussing a rock, but in formal descriptions state grain sizes in millimetres. If a rock is porphyritic, remember to quote the size of the phenocrysts, too; a phenocryst 10 mm long may appear to be 'large' in a fine-grained rock, but not in a coarse one.

6.5.2 Igneous mineralogy

When naming a rock, identify the principal minerals and estimate their relative abundances, using the chart (Appendix III) inside the back flap. Without a chart, you will almost certainly overestimate the quantity of dark minerals by a factor of up to two. Look at a *selection* of grains of every mineral present. identify each mineral in turn, using your handlens. Note the relationships between them. Rotate the specimen in the light to catch reflections from polysynthetic twinning in plagioclases—it is remarkable how many geologists have never seen this except under a microscope. Dark minerals are the most difficult to identify in hand specimen and pyroxene, amphibole, epidote and tourmaline are easily confused. The different cross-sections and cleavages in pyroxene and amphibole should be known to all geologists. Note also that cleavage in amphibole is much better than in pyroxene, epidote has only one cleavage, and tourmaline virtually none. Refer to Dietrich and Skinner (1979) to name these rocks.

6.6 Aphanitic igneous rocks

Aphanitic igneous rocks are difficult to name. Hard and compact, at first sight they appear to give little indication of their identity. Divide them into light-coloured aphanites ranging up to medium red, brown, green and purple; and darker aphanites covering colours up to black. Use the old term *felsite* for the first group and *mafite*, for the second (Dietrich and Skinner 1979). Table 6.1 shows how the groups divide.

Table 6.1

Felsites	**Mafites**
Rhyolites	Andesites (a few)
Dacites	Basalts
Trachytes	Picrites
Andesites (most)	Tephrites
Phonolites	Basanites
Latites (trachy-andesites)	

Adapted from Dietrich and Skinner (1979)

Careful examination of aphanites under a handlens usually gives some pointers to identity and many contain phenocrysts. Basalt is by far the commonest of all black aphanites. In the field, revert to 'spotted black rock' type of terminology if all else fails.

6.7 Quartz veins and pegmatites

Quartz veins are very common and should give no trouble in identification. Some are deposited by hydrothermal solutions along fractures and may show coarsely zoned structures and, sometimes, crystal-lined vughs. Others have been formed by replacement of rock. Some show 'ghosts' of the replaced rock with structures still parallel to those in the walls. Some veins are clearly emplaced on faults; some enclose breccia fragments. Some veins contain barite and fluorspar in addition to quartz, and may even be sulphide-bearing: all veins should be looked at carefully for traces of ore minerals. However, not all veins are quartz veins. Some contain calcite, dolomite, ankerite or siderite, or mixtures of them, and may be mineralized too. Veins by no means always have igneous associations.

Pegmatites always have igneous associations. They are usually, but not exclusively, of granitoid composition. 'Grain-size' may be from 10 mm upwards to over a metre. 'Granite pegmatites' fall into two main groups, *simple* and *complex.* Simple pegmatites are usually vein-like bodies consisting of coarse-textured quartz, microcline, albite, muscovite, sometimes biotite, and rarely, hornblende. Complex pegmatites can be huge with several zones of different composition around a core of massive quartz. They may be mineralized.

6.8 Igneous rocks in general

Always examine an igneous contact thoroughly. Look at both sides carefully and make sure that it is not an unconformity. Note any alteration and *measure* its extent: a 'narrow' chill zone means little to the reader of your report. Sketch contact zones and sample them. Contact metamorphism converts mudstones to hornfelses, hard dense, fine-grained rocks, often spotted with aluminosilicates. They can be difficult to identify. Map them as appropriate, e.g. *grey hornfels*, or *spotted black hornfels*, or *garnetiferous green hornfels*. Sandstones are metamorphosed to quartzites near contacts. Carbonate rocks become *tactites*, or *skarns*; diverse mixtures of calc-silicate minerals. Search skarns with special care for ore minerals, for they are very susceptible to mineralization. Examine contacts between lavas, and the rocks both above and below them, closely, and do not forget the contacts between individual flows.

Few intrusions are homogeneous, yet many maps give that impression because intrusions are so often shown in only one pattern or colour. Map the interior zones of intrusions with the same care you would give to an equivalent area of sedimentary rock. Boundaries between phases may be irregular, gradational, and seldom exposed, but differences in mineral composition and texture, and very often flow banding, can be seen if looked for. Map them. Map also all dykes and veins in intrusions, and record joint patterns.

6.9 Pyroclastic rocks

Treat pyroclastics as sedimentary rocks and apply the same rules when mapping them. They are important markers in geological sequences because they may be deposited over wide areas during relatively short periods of time. Pyroclastic materials are essentiallly glassy ashes. Unconsolidated, they are called *tephra*, when consolidated, *tuff*. *Agglomer-*

ates contain fragments larger than 64 mm, *lapilli tuff* 2–64 mm, and *ashy tuff* anything less than 2 mm. *Welded tuffs* are those in which the ashy fragments fused during deposition. *Ignimbrite* is a special name reserved *only* for rhyolitic welded tuffs. Name tuffs, where possible, for their related lava, e.g., *andesite tuff*, or *ashy andesite tuff*, but many finer-grained varieties are difficult to identify and non-committal names are justified. Some are so glassy, some even flow-banded, that they can be mistaken for lavas in the field. Tuffs tend to devitrify to give spherulitic and perlitic textures. Many weather easily to industrially useful products, such as *bentonite* and *perlite*.

6.10 Metamorphic rocks

Contact metamorphism has been dealt with under igneous rocks. Here we are concerned only with rocks resulting from *regional* metamorphism. Two factors need to be considered when mapping metamorphic rocks: the original lithology/stratigraphy, and present lithology. Whenever possible map them separately.

6.10.1 Naming metamorphic rocks

Sedimentary rocks change with increasing metamorphism, first to slates, then to phyllites, schists and gneisses. Igneous rocks deform and recrystallize to gneisses or schists and many basic igneous rocks, including volcanics, become *amphibolites*.

Name *slates* for their colour, such as brown, green, grey, blue or purple; and for their recognizable minerals, e.g., *pyritic black slate* or *green chiastolite slate*. *Phyllites* cleave more readily than slates, leaving lustrous faces shining with sericite scales.

Geologists will seldom agree where to put the boundary between phyllites and schists in the field. The division tends to be subjective. In general, if *individual* mica or chlorite flakes can be clearly seen, call it a schist, if not, it is a phyllite. *Mica schist* is a common 'sack name'. Where possible define 'mica schists' as *chlorite schist, muscovite schist, biotite-garnet schist*, etc., but not all schists are micaceous, there are *actinolite schists, tremolite schists*, and many others. Unfortunately, schists weather easily and are often poorly exposed.

Gneisses are medium to coarse-grained foliated rocks in which bands and lenses of different mineral composition alternate. Some gneisses split roughly parallel to foliation owing to the alignment of platy minerals, such as micas; others do not. Always qualify the word *gneiss* by a compositional name when first used. As with all other rocks, a locality name can be used as a prefix, or even a more general name, such as *Lewisian gneiss*, to denote gneisses of a certain age.

Gneisses may also be named for their textures, such as *banded gneiss*. Some may contain apparent 'phenocrysts'. These may be cataclased *augen*, or they may be *porphyroblasts* of large new crystals growing in the rock, perhaps replacing former augen. You probably cannot tell which except in thin section, but *augen gneiss* is a convenient field name in either case, even if not always strictly correct.

Migmatites are, literally, mixed rocks. They contain mixtures of schistose, gneissose and igneous-looking material. Treat them in the same way as other gneisses: name them for composition, texture and structure.

6.10.2 *Contacts*

Contacts between many metamorphic rocks are just as sharp as those between most sedimentary or igneous rocks. Some, however, may be gradational, especially within schists and gneisses. Identify every exposure compositionally when mapping them so that gradational boundaries can be inferred where necessary.

6.10.3 *Foliation*

Where structure is fairly regular, map cleavage, schistosity and other foliations at much the same density as for sedimentary rocks. If structure becomes so complex that it is impossible to show it adequately on your map, map it at a larger scale, or make numerous sketch maps and notebook diagrams. A map cluttered with tightly-crowded clusters of symbols is difficult to interpret by its author, let alone by those who may have to refer to it later on.

In addition to foliation, there are many other structures which must be mapped in metamorphic rocks. These include the trend and plunge of any minor folds, whether in bedding, cleavage or other foliations, or even in ptygmatic veins. The sense of folding should be noted, too, to indicate where the major fold closures lie. Style of folding is also important. Look for lineations, including intersection of planar features, such as bedding/cleavage, cleavage/cleavage, etc.; or mineral alignments, rodding, mullions and stretched conglomerate pebbles (Fig. 5.12). In fact, map *any* structure, even if you do not know its significance at the time. Its meaning may become clearer later, or it may not, but at least you have it on record if it does.

6.11 Economic geology

Any geologist worth his salt should at least be able to recognize the principal economic minerals and rocks, for it is his duty to consider the economic, as well as the purely scientific, aspects of any area he maps. To ignore them, or consider them beneath one's scientific dignity—as some do and freely admit—is intellectual snobbery. Before going into the field, review any literature concerning minerals in the region you are to map, both metalliferous and industrial. Note records of quarries and mines. Find out what ores were mined and, particularly, whether they were associated with sulphides, for those ores have distinctive outcrops. Also note the rocks the ores were associated with and keep them in mind when mapping.

6.11.1 *Types of body*

Ore bodies do not necessarily crop out at the surface in easily recognizable form. Some are just rock in which metallic minerals are disseminated; often sparsely disseminated at that. Some *stratiform* zinc-lead ores are merely shales with finely dispersed zinc and lead sulphides, similar in grain-size to the rock minerals themselves. *Porphyry copper* deposits—those large stock-like granitoid intrusions which supply more than half the world's copper—contain less than one per cent metal, and look much like any other intrusion. Take nothing for granted.

6.11.2 *Oxidation*

Ore bodies do not stand up out of the ground with fresh shining crystals of metallic minerals glinting in the sun. Sulphides, especially, are often extensively altered above the water-table by oxidation. Some oxidize to a highly soluble state—copper, zinc and silver ores are examples—and the metals are leached downwards to re-deposit near the water table as a zone of *supergene enrichment*, leaving the upper part of the body depleted (Fig. 6.4). The insoluble iron oxides remaining are left to accumulate at the surface during erosion as outcrops of hard cellular limonite called *gossans*, or as iron-stained soils spread over a wide area.

Rocks near outcrops of ore may be stained by the brightly-coloured basic copper carbonates *malachite* and *azurite*, or coated with the tiny green crystals of the lead chloro-phosphate *pyromorphite*, which are easily mistaken for moss. Blue, green, yellow, red and orange stains should always be looked at carefully.

6.11.3 *Structural control*

Particular attention must be paid to the fracture pattern in any mineral-bearing area, for ore deposition is often controlled by faults and joints. But ore may also be controlled by folds, bedding planes, unconformities, lithological changes, and by contacts where granites and diorites have intruded limestones or dolomites. Ore bodies can be any shape. Some are vein-like, some irregular masses grading into their host-rocks; others are merely an ore-bearing part of an otherwise barren rock—sedimentary, metamorphic, or igneous—and these are the most easily missed.

6.11.4 *Grades and economics*

Before going into the field, study the economics of the rocks and ores you may encounter. For instance, what now constitutes a useful slate? What is a cement limestone? What is a good brick, china or ball clay? What is viable iron ore? Examine specimens of any ores you may expect to see and their oxidized products before you leave for the field. Remember, too, that in the field ores are iron-stained and covered with dirt and may not resemble those scrubbed, and often unrepresentative, specimens seen in collections.

6.11.5 *Water*

Water has been described as the 'essential mineral' and geologists in many countries spend a considerable part of their time looking for it. Much of the search for water is geological common sense. Note its occurrence in any area you map and learn from it.

6.11.6 *Industrial minerals*

Many of the materials you map have a use. That includes many clays, sands and gravels, fluxing materials, even aggregates, roadstones, ballasts and crushed rocks. The variety is immense. Build up your background knowledge of industrial minerals. Bates (1969) and Robertson (1961) are excellent guides.

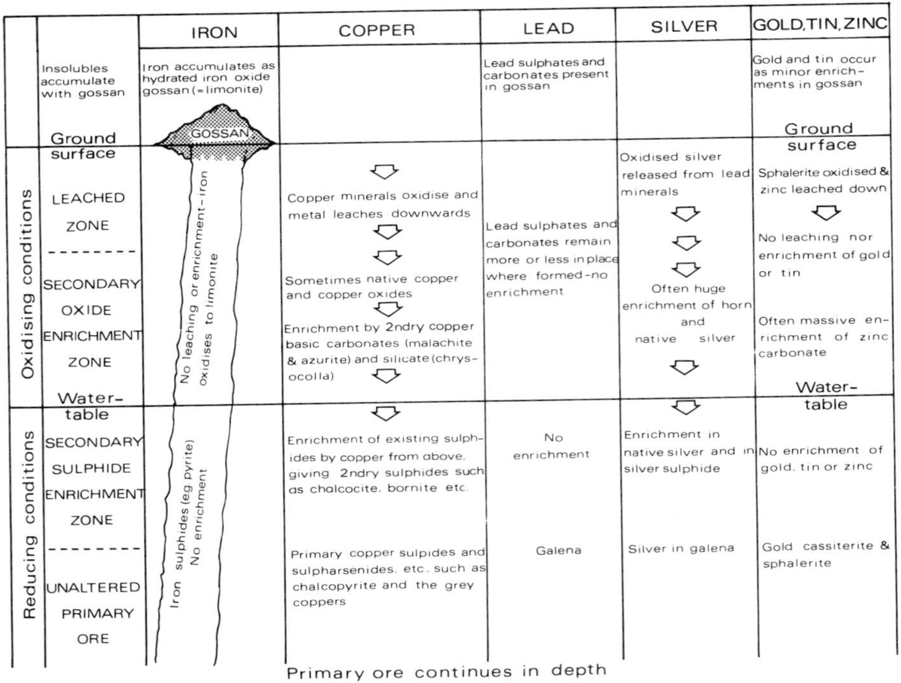

		IRON	COPPER	LEAD	SILVER	GOLD, TIN, ZINC
	Insolubles accumulate with gossan	Iron accumulates as hydrated iron oxide gossan (= limonite) GOSSAN		Lead sulphates and carbonates present in gossan		Gold and tin occur as minor enrichments in gossan
	Ground surface					Ground surface
Oxidising conditions	LEACHED ZONE	No leaching or enrichment – iron oxidises to limonite	Copper minerals oxidise and metal leaches downwards	Lead sulphates and carbonates remain more or less in place where formed – no enrichment	Oxidised silver released from lead minerals	Sphalerite oxidised & zinc leached down No leaching nor enrichment of gold or tin
	SECONDARY OXIDE ENRICHMENT ZONE		Sometimes native copper and copper oxides Enrichment by 2ndry copper basic carbonates (malachite & azurite) and silicate (chrysocolla)		Often huge enrichment of horn and native silver	Often massive enrichment of zinc carbonate
	Water-table					Water-table
Reducing conditions	SECONDARY SULPHIDE ENRICHMENT ZONE	Iron sulphides (eg. pyrite) No enrichment	Enrichment of existing sulphides by copper from above, giving 2ndry sulphides such as chalcocite, bornite etc.	No enrichment	Enrichment in native silver and in silver sulphide	No enrichment of gold, tin or zinc
	UNALTERED PRIMARY ORE		Primary copper sulpides and sulpharsenides, etc., such as chalcopyrite and the grey coppers	Galena	Silver in galena	Gold cassiterite & sphalerite

Primary ore continues in depth

Fig. 6.4 The oxidation of deposits containing iron, copper, lead, silver, gold, tin and zinc. (First published in *Subterranean Britain* (ed. Crawford) by A. & C. Black (Publishers) Ltd.)

7
Field maps and field notebooks

Field maps and notebooks are valuable documents which constitute part of the record of the field evidence on which the interpretation of geology depends. Both are the property of your employers and will be retained by them as part of their permanent records when you leave them. The reason is obvious. If your erstwhile employer wishes to reinvestigate an area you mapped, then it will be necessary to refer back to the original field records.

7.1 Field maps

7.1.1 Data needed

A field map is an aid to the systematic collection of geological data in the field and shows the evidence on which the interpretation of the geology was made. It shows the geological features you actually *saw* in the field: it also shows geology you have *inferred* from indirect evidence, such as changes in topography or vegetation, spring lines or float. A field map is not an interpretative map as such, but all contacts should be plotted on it in the field, though some may have to be inferred from minor indirect evidence or sometimes merely by your judgement of where they most probably occur. However, fact must always be clearly distinguished from inference. A field map is not merely a rough worksheet on which to temporarily plot information before transferring it on to a 'fair copy' map back at camp or base; it is a valuable research document which you or others may later wish to refer to. No evidence should be erased from it to 'tidy it up', or because it is not needed to aid the present interpretation, nor should you add anything to it at a later date which you think you saw in the field but did not record at the time. The type of informatin to be recorded on a field map is:

1. The location of all rock exposures examined.
2. Brief notes on the rocks seen.
3. Structural symbols and measurements, such as those for dip and strike.
4. Locations to which more detailed notes in your notebook refer.
5. The location from which each rock or fossil specimen was collected.
6. The location at which every photograph was taken or field sketch made.
7. Topographic features from which geology may be indirectly inferred but which are not already printed on the map. Changes of

slope or vegetation and the positions of seeps and spring lines are examples.

8. All major contacts, including faults, both certain and inferred.
9. River terraces, beach terraces, and similar features.
10. Alluvium, scree, boulder-clay and any superficial materials, including landslide debris.
11. Cuttings, quarries and other man-made excavations exposing geology eg. pits and boreholes.
12. Comments on the degree of exposure or lack of exposure, and on soil cover.

Because they are valuable, field maps should, as far as possible, be kept clean and protected from rain and damp. This is not always possible and important information must not remain unplotted for fear that the map may get wet or dirty if the map case is open in the rain.

7.1.2 Preparation

Before using a new map sheet, cut it into a number of sections or 'field slips' which will fit into your map case without having to be folded. Folding ruins a map: it is difficult to plot any information close to the folded edges (especially if folded over twice) and any information which is plotted there is soon smeared and eventually worn off. Every field slip should be titled, have a scale, and carry a full explanation of the colours used unless they follow the strict conventions of the organization you are working for. Any non-standard or unusual symbols used should also be shown, together with a diagram of how the several field slips which make up the whole sheet are numbered and relate to each other. The number of the notebook which refers to the slip should also be included, together with the name of the author of the map and the dates of starting and completing it. Write this information on the reverse of the field slip, (Fig. 7.1). On the face-side of the map the north direction from which structural measurements have been plotted should be shown; true, grid, or magnetic north as the case may be.

Do not stick your field slips together with adhesive tape when fieldwork is complete. It makes them awkward to use again in the field if new information has to be added, and most self-adhesive tapes shrink with time, dry out, or come adrift leaving a dirty stain.

7.1.3 How and what to plot

A field map is a record of field observations of the type listed at Section 7.1.1. Plot the position of exposures seen and indicate rock type by formation letters, letter symbols, or by colouring, supported where necessary, by notes on the rock condition. Keep notes short and use abbreviations such as fg (for fine-grained), lam (laminated), shd (sheared). Refer also to Section 4.3.1. Many exposures need no more than an outline drawn to show their limits, shaded with the appropriate coloured pencil. Exposures which are so small that they can only be shown by a coloured dot should, however, always be supported by a letter symbol, otherwise they tend to be overlooked when inking-in the map later. If notebook notes are made at an exposure, then the location on the map must be linked to the notebook record (see Section 7.1.5). Structural observations are shown by

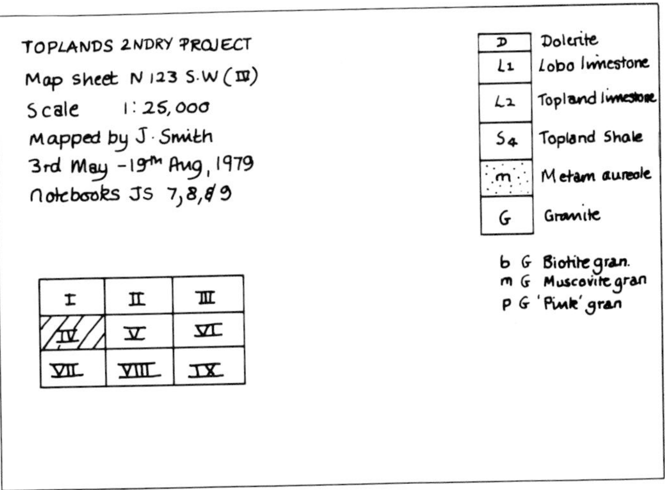

Fig. 7.1 Reverse side of a field slip with the information it should carry. Note the index showing how this slip relates to the others which make up the whole field sheet.

the appropriate symbols, drawn large enough to enable them to be traced off accurately on to the fair copy on a light-table: 6–7 mm is a suitable length for strike symbols. Print the numerical value of dip or plunge legibly in such a position that there is no ambiguity as to which symbol the figures refer to. Even better, record both dip and strike by using the 'right hand rule' (Fig. 7.2). It is unnecessary to enter every dip and strike, trend and plunge, in your notebook, providing you draw your symbols large enough on the map itself and record strike/dip etc. in figures beside them. Booking unnecessary information merely consumes time better spent on mapping.

Contacts should be shown as continuous lines where seen on the ground, with a note or symbol to indicate their type. Do not try to distinguish between different types of contacts by different *pencil* line thicknesses. Distinguish faults by the letter 'f' or, if the dip is known, by a dip arrow (see list of symbols in front flap). Inferred contacts are shown by broken lines and different types of inferred contacts can be distinguished by the frequency of the breaks. Keep the breaks in broken ('pecked') lines small, otherwise the lines look untidy. Show thrusts with the traditional 'saw teeth' on the *upper plate*, but do not try to draw the teeth as closely as those on printed maps: a tooth every 1–2 cm is quite adequate for a field map, and if you do make a mistake, is far easier to erase after inking-in (Fig. 7.2).

Although a field map is essentially a factual data map, this does not prevent you from plotting the inferred positions of contacts deduced from

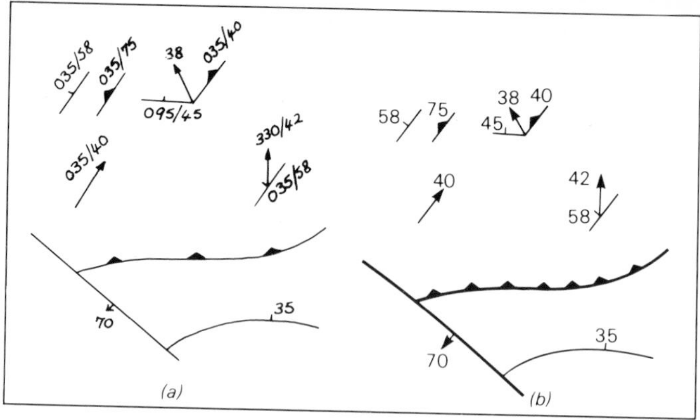

Fig. 7.2 Symbols for field (a) and fair copy (b) maps compared. Strike and trend, dip and plunge, are shown on the field map, but only dip and plunge on the fair copy.

indirect evidence such as vegetation, spring lines and breaks of slope. In fact, the field is the proper place to infer contacts, for there is usually some evidence, however slim, of their positions. The drawing of contacts in the office, or back in camp, is only justified if you have to resort to geometrical constructions owing to a complete lack of evidence on the ground, or when you have geophysical or photogeological information to help you.

Any topographic features which may reflect concealed geology but are not already printed on the base map should be added to it. These include breaks of slope, vegetational changes, distinctive soils, springs and swampy patches. Show also landslides, scree and alluvial terraces. Outline mine tips because they can often provide fresh specimens of materials which are otherwise seen at the surface only in a heavily weathered state—this is particularly so in humid tropical climates—or even prove that certain rocks which are not exposed, or are quite unsuspected in the area, do occur at depth.

The amount of detail that can be shown on a map obviously depends on its scale. A field slip should not be so cluttered with information that 'the wood cannot be seen for the trees', but even more difficult to interpret is the map which shows almost nothing but a series of numbers referring to notebook entries. There is a happy medium between these extremes. The face of the map should contain all relevant *basic* geological information: the notebook should expand it and provide details of features which are too small to show on the map. The complexity of the geology and its degree of exposure also, of course, determines how much can be shown. Sometimes a small scale may be deliberately chosen to restrict the amount of detail in reconnaissance surveys when a larger scale might tempt the geologist to spend too much time on details.

On the other hand, if the main object of the work is to solve specific geological problems, then the scale must be large enough to show, without crowding, the type of detail that must be mapped to solve those problems; and if maps of a suitable scale are not available, then they may have to be made. Often, however, complexity of geology and degree of exposure vary from one part of a region to another so that very large-scale maps need only be made over limited areas, with considerable saving in cost. Frequently, the results of small-scale mapping indicate areas which require re-mapping on a larger scale: this is particularly so in mineral exploration where larger and larger scale maps of smaller and smaller areas may be made as the more interesting localities are recognized and unmineralized ground eliminated. If *occasionally* you are forced to make more extensive annotations on a map than space will allow, then make a small needle hole at the locality, and write your notes on the back of the map—if not already written on—but do not make a general practice of it. If you need more space: use a larger scale.

7.1.4 Neatness

Information written on the map must be written as legibly as circumstances allow. Keep one pencil, and a reasonably hard one at that, for plotting on your map, and another for your notebook. Keep your plotting pencil needle sharp, otherwise you cannot plot or write legibly on your map. If you use only one pencil, you will be continually sharpening it between notetaking and plotting. Write on your map in a fine, clear *printed* script. Do not use a miniature cursive running handwriting; this is far less legible, especially when written under the often difficult conditions of fieldwork. Do not use stylus-type pens in the field: everyone makes mistakes and they are far more difficult to remedy if made in waterproof drawing ink; secondly, notes frequently have to be erased and rewritten because they overlap some geological feature you had not found when you first made them. Even when inking-in pencil-written notes, you frequently have to rearrange them so that they are neater, more legible and parallel to one another. Drawing and draftsmanship is an essential skill for any field geologist: if he cannot draw neatly, he cannot map accurately. Much of the skill needed can be acquired by effort and practice.

7.1.5 Linkage of map localities to notebooks

The most practical permanent way to link map localities to notes in your notebook is to use map (grid) references (Section 3.3.2). Map references have the advantage that points can be located by a group of figures with great accuracy and without ambiguity and, even if the original field slip is lost, the points can be relocated on any map of any scale covering that area. In general, however, relocating map references on your map during interpretation is slow and irritating. Easier to use is the simple consecutive numbering of observations. This works well, providing the points on the map are fairly closely spaced along more or less specific directions, such as traverse lines. Consecutive numbering is improved by designating each grid square printed on the map by a letter, or by the map reference of the SW corner of each grid square, and then giving consecutive numbers to the observations within those squares. Whatever you do, always enclose observation numbers on the map by a circle to avoid confusing them with dip readings. Notebook entries are numbered A1, A2 ... A23, or 8746/1 ... 8746/2 (Fig. 3.1), etc., and there is seldom any difficulty in rapidly relocating them on the map, but always draw a diagram in the front of your notebook to show later readers how the letter symbols relate to map squares. The drawback is that if you lose the field map the notebook becomes virtually useless.

Map localities can also be identified by notebook page number. If several notes are made on the same page, designate them a, b, c. When more than one book is used for a project, prefix the page number with the book number. Locality 5/23b, for example, means note b on page 23 of notebook number 5.

7.1.6 *Inking and colouring field slips*

Observations made on field slips during the day should be inked-in at night. Even on the best-protected maps, fine pencil lines become blurred or lost with time. When 'green line' mapping, exposure margins should be outlined with green *waterproof* ink or, in sunnier climates, with a fine closely-dotted line in Indian ink. After inking, recolour each exposure with the appropriate coloured pencil. Ink traverse lines with a continuous line where geology is certain and a broken line where inferred, then overlay the line—continuous or broken—with a coloured pencil line.

Inked contacts can be distinguished by lines of different thickness drawn with stylus-type pens, but notes should still be added to confirm their characters: abbreviations, such as 'f' for fault and 'uc' for unconformity are sufficient. Unmarked contacts are presumed normal.

Ink all structural symbols and rewrite the amounts of strike/dip etc. Rewrite notes in fine neat script so that they cover no geological features and align them so that all, as far as possible, are parallel with the same direction. It is irritating having to turn a field slip first one way, then another, to read the information on it.

There may well be interpretative lines shown on a field slip at the end of day whose positions are still uncertain: leave them uninked until you are quite certain of their validity, even it it means repencilling their traces each evening to avoid losing them. Add any information from air-photographs to your map in waterproof ink—purple for general geology, red for faults—to distinguish it from information mapped on the ground. This discrimination in no way diminishes the validity of photogeological information; but it does distinguish your sources of information and also shows you where features can be confirmed on the ground.

Having inked in your map and reviewed your day's work, then, if you must, *lightly* shade or cross-hatch in colour those areas which you now infer are underlain by specific rock types. DO NOT colour in your map heavily, as if it were a 'fair copy' map. DO, however, re-colour your traverse lines and areas of exposure more strongly, so that they stand out as the evidence which justifies your interpretation. Geologists only too often map exposure by exposure during the day, carefully distinguishing what they have observed from what they have inferred, only to obliterate all their field evidence in the evening by swamping the map with solid colour in an endeavour to make it look like a finished geological map. A field map is an 'evidence' map. It is not a rougher version of a fair copy, and should not look like one. Make sure that fact can be distinguished from inference on it.

7.2 Field notebooks

Like field maps, field notebooks are valuable documents that form part of the record of field evidence on which the interpretation of geology depends. A field notebook will be referred to by workers who reinvestigate the area it concerns at least as often as the field map it relates to, perhaps to elucidate data on the map, perhaps to obtain details of specimens or fossils collected. Later workers may want further details of specific exposures or lithological sections to discover why you drew the conclusions you did, or your notebooks may provide information which is no longer available: exposures may be built on or dug away, pits and quarries may be filled, or records destroyed. Notebooks must therefore be kept in a manner that others can understand and, above all, they must be legible. The US Geological Survey insists that their notebooks are written only in hand-printed script. This helps to make even those notes written with ice-cold hands on a wet and windy day more legible. Sketches and diagrams too must be properly drawn and labelled, dimensions given, and where appropriate, tinted by coloured pencil.

Develop a habit of *using* your notebook. During a project, non-geological records, such as of expenses, also have to be kept. Where better than in your field notebook? Use your notebook as a diary and even if no work is done on a particular day, such as a Sunday, record the fact. Even visits to cinema or pub can be noted, as social events can often help you to remember more geological occurrences that you saw on the same day. Do remember, however, that others may read your notebook later! Only too often notebooks suffer the fate of diaries. Copious neat notes are written on the first day, fewer, rougher ones on the second. By the end of the week, notes are sketchy, untidy and illegible. Your field notebook is as important as your field map. Use it properly.

7.2.1 Preliminaries

Write the name of the project, the year, and the notebook number, on the cover of every book. Inside the cover, write your own name and address in waterproof ink, and offer a reward for its return if found. Be generous because the loss of a notebook can be disastrous. If necessary, repeat the information in the language of the country you are working in. Number notebook pages, but leave the first few free for an index to be filled in, day by day, with a list of contents and their page numbers. This helps not only others who may have to use your notebook, but also you yourself when you come to look up information for your report from notes made months, perhaps even a year or more, before. Remember that

(45)

of Haem on lwr. contact

Dark green porph also intrudes the dol. & there appears also to be mineralisation. (in float) but no solid Fe exposure

A41 Haem.
A42 Skarnified porph.
A43 Haem.

26th Aug. Mon.

Traverse across sequence: at 'Miocene Ridge'

'Miocene' Ridge

Westwards along base of dip slope small 'flat-irons' of thin limey sst interc. with sh, overlying basal Ls. These sandy beds are conglomeratic at base & more limey in lower parts. They are probably lenticular & thin rapidly. The interc. sh also show great variations in thickness. Thin beds tend to sep E'wards you keep a notebook to refer to, so make it easy to find things in it. Ask yourself 'What use would this notebook be to me if I had to refer to it a year, two years, or even five years from now?'. If you doubt that you could understand it yourself, no one else will be able to.

Add also to your notebook, registers of rock specimens, fossil specimens and photographs. Tape on to the last few pages xeroxed tables of pace lengths and charts such as the 'percentage area' chart given in Appendix III and fasten a piece of fine sandpaper into the back cover too, to keep your pencils sharp.

7.2.2 Linking notes to map localities

Methods to link observations made on your map to your notebook are given in Section 7.1.5. Write the map references or note numbers in a column on the left-hand side of the notebook page. Use this column only for note, specimen and photograph numbers. Write note numbers in pencil, specimen and photo numbers in red and blue pencil, respectively, so that they can be quickly spotted. If numerous, give specimens and photos a column of their own.

7.2.3 Recording information

The purpose of a field notebook is to expand information on your field map, not to duplicate it. For instance, normally there is little point in writing down the numerical values of dips and strikes plotted on the map unless weather conditions are so bad that you cannot plot them in the field. If for any reason dips and strikes have to be recorded—joints are an example—then make the information easier to retrieve by recording it on the right-hand side of the page. Write your notes as briefly as possible, even omitting verbs at times, providing you do not lose the sense of meaning. Use abbreviations where appropriate: there are many all geologists understand, such as ls for limestone, sst for sandstone, and sch for schist, or flt for fault and jts for joints. Tabulate in the front of the book any non-standard abbreviations you use unless their meaning is obvious. Use any short cuts in the field which save time without loss of information.

7.2.4 Sketches

Use sketches to supplement notebook descriptions whenever possible. Sketches should show dimensions, or at the very least, some indication of scale. Ink-in complex sketches, especially those which carry fine lines, lettering and dimensions, to avoid loss of details. The practice of inking-in notebook notes in general, however, is quite unnecessary. A specimen page from an actual notebook is shown in Fig. 7.3.

7.2.5 Cross-sections

The understanding of the structure of an area is aided by plotting cross-sections in your notebook along selected

Fig. 7.3 Page from a field notebook. The column on the left shows the registered number of specimens collected. The observation number referring to the locality appears in the same column on the previous notebook page. The cross-section on the lower part of the page needs no locality/observation number, provided section lines are shown on your field slips.

lines in the field (Section 4.1.2). Their presence in a notebook is most useful to anybody reviewing the geology of the region later and show the validity of a structural interpretation far better than those beautifully drawn and coloured cross-sections which accompany redrawn ‘fair copies’.

8

Fair copy maps and other illustrations

8.1 Fair copy maps

Geological field maps are records of factual observations made in the field: they are not interpretative maps. Therefore, when mapping has been completed, you must compile a 'fair copy' manuscript (i.e., hand-drawn) map from your field maps, notes, and your later laboratory work, to accompany your report. The fair copy is not merely a redrawn version of the field map: geological formations are now shown as continuous units instead of disconnected outcrops. It is also a selective map and it may well be that some formations distinguished on the field slips are not differentiated when boundaries are transferred to the fair copy. This may be because the distinctions made in the field were found to be geologically less important than first thought, or because some units were so discontinuous that they could only be traced over very short distances. Alluvium, swamp, peat and bog are shown on the fair copy, as also are laterite and boulder clay, but soils are not. The general rule is to show any features which add to the understanding of the geology and to omit those which do not.

Much of the information gathered during field work is not transferred to the fair copy. For instance, the notes made on the face of the map are not normally shown although generalized comments may be made, such as 'cordierite schist' if this zone is not indicated by a specific symbol of its own, or 'red soils' to justify the continuation of an unexposed dolerite dyke. More specific notes, such as 'malachite stains' may occasionally be needed, but otherwise notes are used only to emphasize specific or unusual details, or justify the geology. Specimen locality numbers should not appear on the fair copy, either, although ore mineral and fossil localities may be marked where their presence is geologically or economically significant. The criterion for what to transfer is really a matter of common sense. The finished fair copy map should show the geology of the region in such a manner that the geological formations can be clearly distinguished one from another, and if they are continuous units it should be possible to trace them from place to place across the map even though poorly exposed on the ground. Structural symbols should be sufficiently clear so that the sequence of events can be elucidated and the stratigraphy determined. Above all, the map should be neatly drawn, the colours smooth and distinctive, and the printing legible.

It may be asked why so much information, painstakingly collected in the field, is omitted from the fair copy. It is because the fair copy is only part of your geological interpretation. It is essentially an index map which provides the basis for understanding any accompanying explanatory report. The map is not an end in itself, but it should still be able to stand on its own, showing the general features of the geology in a clear and concise manner.

8.2 Transferring topography

A fair copy map is usually drawn on a new copy of the original topographic base map used in the field. If the fair copy has to be made on tracing paper or film, then sufficient topography must be traced off the field map to make the geology understandable. This is tedious, but it is necessary. In very mountainous districts sufficient relief may sometimes be shown by tracing every second, or even fifth, contour. Draw contours in brown unless 'dye line' (ozalid) copies are to be made. In that case, as brown ink reproduces patchily, show them in black as broken lines or with a thickness that can be clearly distinguished from geological boundaries. Sometimes, fair copies have to be drawn on tracing paper or film because the printed base map is so cluttered with coloured geographic detail that the geology would be lost if plotted on top of it. This is particularly so on maps of some of the more mountainous parts of the world. Some maps drawn on tracing film are made from aerial photographs. In this case trace off main drainage and mark hilltops, at the very least.

8.3 Transferring geology

Usually, when preparing a fair copy map, information has to be transferred from field slips on to a clean, new opaque paper base map. There are several ways of doing it. The first is to copy information purely by inspection. This can only be done if there is sufficient printed detail on the map to serve as reference points. If there is not, then pencil a grid in on both field slips and base map and transfer the information square by square. In either case, strike symbols must be replotted from the original data. A better way is to use a *light-table* so that detail can be traced from field slips directly onto the base map. If the printed geographic detail on the well-used field slips does not exactly fit that on the brand new fair copy base sheet owing to scale changes caused by weathering, you must constantly adjust each field slip beneath the fair copy to maintain as good a fit as possible as detail is traced off it. A light-table is not difficult to make. Sink a window of opalescent perspex, 20–30 cm square, into a large piece of plywood mounted on a well-ventilated box. Light the window from below by two short fluorescent tubes. To use, move the working area of the map over the window as required.

8.4 Lettering and symbols

Bad printing can ruin an otherwise well-drawn map while good lettering can improve even a poorly drafted one. 'Transfer lettering' is one solution but is expensive and better kept for titles and subtitles. Stencils cannot produce such a fine result as transfers, but can be used over and over again. Mistakes, however, are more difficult to remedy. For small-sized printing, such as descriptions of rock units in the map explanation, for notes and formation letters on the face of the map, and for ancillary explanatory notes, stencils are probably quicker and easier too.

Despite such aids as transfer lettering and stencils, every geologist should develop a good legible hand-lettering style, for there will be many occasions in his life when this is the only way he can letter his map: inking-in a field map is a good way to get practice. *Italic* letters are easier to print than upright, and always use parallel guide lines except for the smallest sizes. Draw all strike symbols on a fair copy exactly the same size: 5 mm long is suggested. Lineation arrows can be a little longer. Draw arrow heads neatly with an ordinary mapping pen. Print figures for dip (bearing of strike is omitted now) *either* parallel to the symbol or parallel to the bottom edge of the map, but not in both directions on the same map (Fig. 7.2). The symbols printed inside the front flap conform to those generally accepted around the world. A much more extensive list compiled by staff of the Australian Bureau of Mineral Resources is given in the *Field Geologists' Manual* (Berkman, 1976, p. 146–64).

8.5 Formation letters

Every rock unit which appears on the face of the map, whether sedimentary, metamorphic, or igneous, must have a distinguishing symbol or 'formation letter(s)' assigned to it. Established formations may already have officially recognised symbols, such as **d^{6a}** for the *Lower Pennant Measures* of the British Carboniferous System, **Mld** for the Mississippian *Leadville Dolomite* (Formation) in the United States and **M2** for the *Upper Red Formation*, better known as the 'Fars', of Iran. If no symbol has been allocated to a formation, you must do this yourself, following the convention of the country, if there is one. Otherwise use the initial letter of the unit wherever possible, as this acts as a mnemonic, and avoid calling your units A, B, C, or 1, 2, 3. Show formation letters on every area of each rock unit that appears on the map. Where a unit covers a very wide area, the formation letters should be repeated several times, but where an area is too small to contain them, print the letters beside it with a 'leader' (⌒) pointing to the area.

8.6 Layout

A fair copy map should be properly laid out. It should have a proper title, a scale, north points and an 'explanation' of the symbols used, together with a record of the authorship of the map and accreditation of any other sources used, including the source of the base map itself. It should also show the date of the fieldwork and the date of its publication. The arrangement of this matter requires some thought and may warrant making a mock-up on tracing paper, or at the very least a rough sketch so that the sheet looks properly balanced. If the map is stuck down to a larger sheet of good quality thin paper, the ancillary information can be arranged around it, or at one side of it, as in Fig. 8.1. Express the scale of the map in both figures, as a 'representative fraction' (i.e., 1:10,000), and graphically, as a 'bar scale' (Fig. 8.2). North points should be as plain as possible so that they are quite unambiguous. Show the differences between true, grid and magnetic norths at exaggerated angles so that there can be no mistake in their relative positions. Print the amounts of the deviations in words and figures beside them, and do not forget the annual rate of magnetic change. In the explanation draw the symbols at exactly the same size as those used on the face of the map: this is not always done.

8.7 Colouring

A manuscript fair copy map is normally hand-coloured by its author before being presented. Colour your map by whatever method you think you are capable of doing neatly and which the paper will take. Many maps are spoilt by this final colouring. Water colours probably give a more finished look to a map but they are very difficult to apply over large areas, especially if those areas have intricate boundaries: irregular drying marks mar the result. Coloured pencils give excellent results with a minimum of practice. Lay on colour gently and with care and, if not dark enough, then add another layer on top of the first. Smooth the colour to

Fig. 8.1 Layout of a fair-copy map, showing the arrangement of explanatory matter. Note the terminations to cross-section lines shown on the face of the map to ensure that there is no ambiguity over where section-lines end.

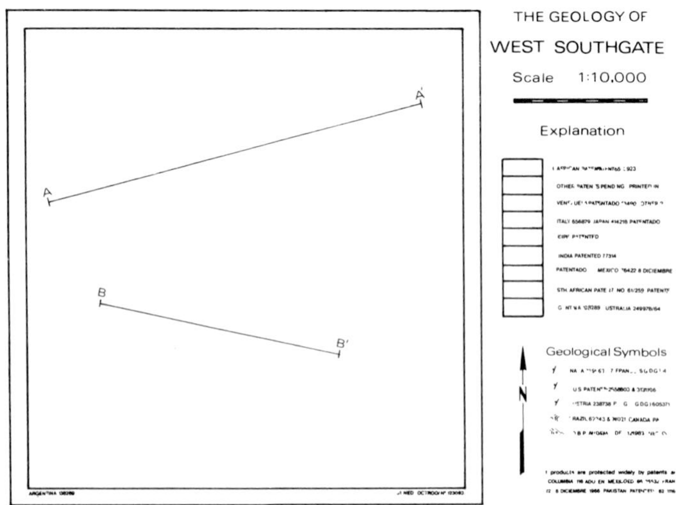

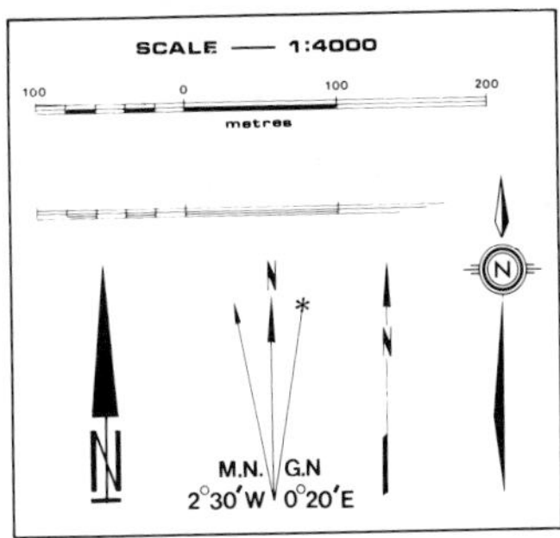

Fig. 8.2 Scales and northpoints for fair-copy maps. Two examples of 'bar scale' are shown and a selection of northpoints. Avoid the 'arty' N-point on the left. The two on the right are printed transfers. A composite northpoint is shown in the centre.

a more even tint by rubbing the surface with a tissue or cotton swab, or by using a 'pastel stump'. Some coloured pencils react better if the swab is first damped with water, petrol or other solvents. Coloured pencils will give patterned effects if a textured surface, such as sandpaper or a book cover, is placed beneath the map when colouring. Different textures of the same colour can be used to distinguish related formations and so extend even a limited range of colours; or darker textured colour can be overlaid on a plain colour. Alternatively, coloured dots can be added to a base colour with a felt-tipped pen. Dots can even cross geological boundaries to indicate, for instance, a thermal aureole.

Choose colours with care and with due consideration for tradition. In general, use pale colours for rock units which cover wide areas and strong colours for rocks with limited outcrop, such as thin beds and narrow dykes. Always keep your reader in mind. Try to follow a system which does not force him to keep looking back to your map's explanation to find out what things mean. Relate your colours to rock mineralogy. For instance, if a hornblende schist is shown in pale green, or overlain with green dots, and a biotite schist is shown in brown, or is overlain with brown dots, then your reader will probably find your map easier to follow than if these formations were shown in purple or blue. Note, however, the use of colour does not obviate the need for formation letters.

8.8 Cross-sections

Display cross-sections in the map margins whenever possible so that all geological information is kept together. If sections have to be drawn on separate sheets, draw all on the same sheet so that they can be easily compared. Show the positions of all sections presented with the fair copy map by lines drawn on the face of the map with each end of each section clearly indicated, as in Figure 8.1; and always draw cross-sections so that their northern and eastern ends are at the right-hand side of the sheet.

Although the horizontal and vertical scales of the section will usually be the same, both a horizontal bar scale and a graduated vertical scale should be provided. Finally, geologists should curb their imagination on fair copy sections and not show interpretations of geology down to improbable depths for which they can have no possible evidence. This does not mean that hypothetical cross-sections should never be drawn, but that their proper place is in your report, with supporting text.

Only factual sections should form part of a factual map.

8.9 Overlays

Do not over-crowd a fair copy map with specialized information in addition to geology, such as rose diagrams, joint measurements and structural statistics. This information is better drawn on transparent paper, or film, as an overlay to the map. You need not limit the number of overlays. In addition to those mentioned above they can include: folding axes; sub-surface contours on specific beds determined from drill holes; isopachytes, isopleths, geochemical contours, and even geophysical information. Not only can overlays be superimposed on the fair copy map, they may be usefully overlaid on each other.

An overlay should be the same size as the fair copy sheet and have the same general format. Show the margins of the map area and, because the map and overlay are of different materials which distort differently, draw 'register' marks to fit the grid intersections on the fair copy.

Title the overlay and give it a simplified bar scale, a north point, and an explanation of the symbols used. Add a subtitle to indicate which map the overlay refers to and the source of any information which does not originate from the authors.

8.10 Text illustrations

Text illustrations are needed in nearly all geological reports. The simplest are merely outline diagrams to explain a single point. Figure 5.9, which shows the 'right-hand rule', is an example. Field sketches redrawn from a notebook are often included in reports. Keep them as simple as possible; show only the salient points. Sometimes sketches can be traced from photographs but again trace off only the principal outlines. Outline sketches are far more comprehensible to the reader than unskilled artistic attempts to reproduce every detail faithfully. Stick-down sheets of stipples and cross-hatching, sold under the names of *Letratone*, *Chartpak* and *Zip-a-Tone*, can enhance your drawing, as in Fig. 5.15.

9
Cross-sections and three-dimensional illustration

No geological map can be considered complete until at least one cross-section has been drawn to show the geology in depth. Cross-sections explain the structure of a region far more clearly than a planimetric map. They may be drawn as adjuncts to your fair copy map, and as simplified text illustrations in your report. In addition to cross-sections, columnar sections can be drawn to show changes in stratigraphy from place to place, or 'fence' or 'panel' diagrams to show these variations in three dimensions. Refinements in three-dimensional illustration include 'block diagrams' which show the structure of the top and two sides of a solid block of ground and models to aid interpretation, such as 'egg-crates'.

9.1 Cross-sections

Cross-sections are either trial sections drawn to solve structural problems, or are drawn to supplement a fair-copy map or illustrate a report.

9.1.1 Trial cross-sections

Draw a cross-section whenever a problem of interpretation arises. Do it when possible while still at your field camp so that you can take additional structural measurements if needed. Even when you have no problems, sections should still be drawn during the fieldwork stage to ensure that nothing is being missed. In geologically complex areas there may be more than one apparent interpretation of the structure and trial cross-sections will at least show which is the most probable. Drawing cross-sections should become second nature to a geologist.

9.1.2 Fair-copy cross-sections

A fair-copy cross-section is drawn to accompany a fair-copy map. Draw it to the same standards, and colour it in the same tints, for it is to all intents and purposes a part of that map. Present it on a separate sheet, with all other sections of the same map, or draw it in the map margins.

Draw cross-sections as if you are looking in a general westerly or northerly direction so that the southern, south-western and western ends of the section always appear on the left-hand side of the page, and the northern, north-eastern and eastern ends, on the right. Draw them to cut across the strike of beds as close to a right angle as possible; if there is a broad swing in strike across the map, change the direction at a few well-separated points to keep it nearly perpendicular to strike. Normally, to avoid distortion, horizontal and vertical scales should be the same but where dips are no more than 10°, an exaggerated vertical scale is permissible, but always state the true dip on the section.

9.1.3 Serial cross-sections

Serial cross-sections are drawn along regularly-spaced parallel lines, usually on large-scale plans used for mining or engineering purposes. They may be drawn at right angles to the strike of the structure, but more usually they are drawn parallel to one set of grid coordinates.

9.1.4 Text figures

Simplified cross-sections are frequently used as text figures to illustrate specific structures described in reports. The vertical scale can be legitimately exaggerated to clarify specific points.

9.2 Plotting and drawing cross-sections

9.2.1 Construction

Poorly-drawn sections are so often encountered in professional life that a résumé of the process is given below:

1. Draw the line of section (A–A′) on the face of the map, marking each end of the line with a short cross-line (Fig. 8.1).
2. Fasten the map to a drawing board with the section line parallel to the bottom edge of the board.
3. Tape to the map, a few centimetres below the section line, a strip of tracing paper on which to plot the section.
4. Draw a base-line on the tracing paper parallel to the section line on the map. Then draw a series of parallel lines at the chosen contour interval above it (Fig. 9.1a).
5. Tape down a plastic rule or steel straight edge so that it cannot move, well below and parallel to the base-line.
6. By sliding a set-square (triangle) along the straight edge, drop a perpendicular down to the appropriate elevation on the section paper from every point where the section line cuts a contour line on the map (Fig. 9.1a). Join these points to give the profile of the topography.
7. Calculate the *apparent dip* in the line of section for every strike/dip intersected by the section line (Section 9.2.2). Mark the position of each strike/dip symbol on the profile and plot the apparent dip as a short line (1–2 cm) (Fig. 9.1b).
8. Project, in its direction of strike until it meets the section line on the map, any strike symbol lying close to the cross-section line, but not actually crossed by it. Calculate its apparent dip, and plot it on the cross-section profile, as before. The distance you may project a strike is a matter of geological judgement. Where there is obvious flexure, extend the strike line to follow its curve to meet the line of section (Fig. 9.1b).
9. Still using the set-square, drop a perpendicular wherever the section line crosses a geological contact on the map and lightly mark the position on the profile of the topography.

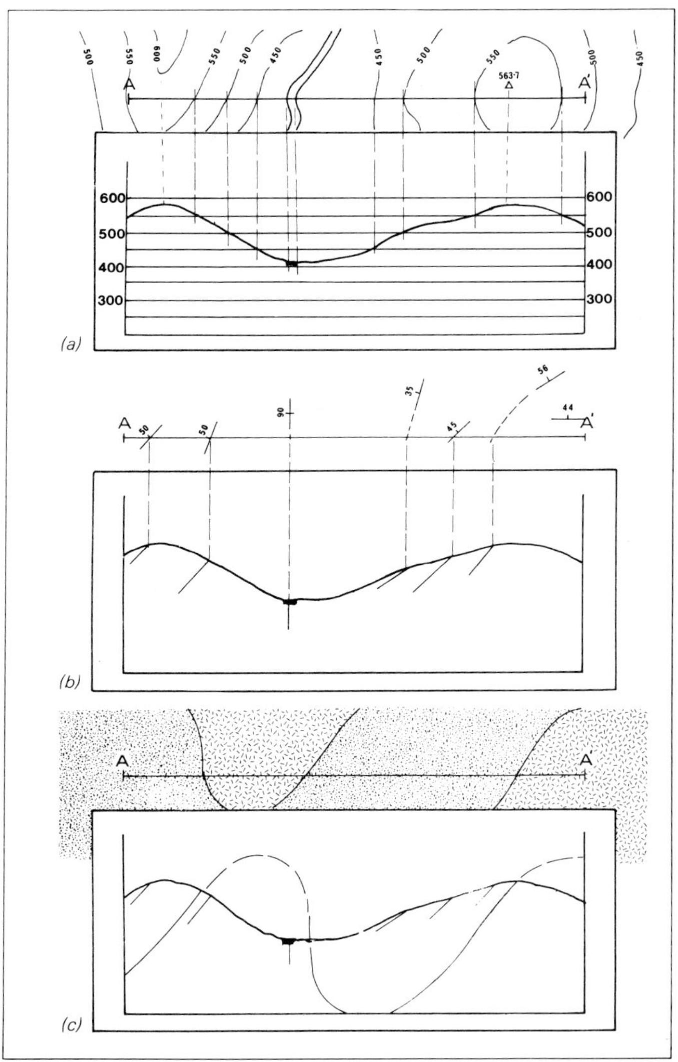

Fig. 9.1 Drawing cross-sections (see text).

10. Lightly sketch in the structure, extending the 'dip lines' and drawing contacts parallel to them. Then modify the interpretation to allow for thickening and thinning of beds, and for any other suspected change in straightforward folding or tilting. Do not interpret , geology to improbable depths beneath the surface. Test your interpretation by continuing the structure above the topographic surface: you have just as much evidence there as you have for below the surface (Fig. 9.1c). Finally, ink-in your interpretation, including, where appropriate, a dashed line to indicate structure now eroded away.

The task is made easier if a 'T-square' or a drafting machine is available. Badgley (1959) describes various ways of drawing cross-sections in many types of structural settings, including geometric constructions such as the Busk method.

9.2.2 *Calculating apparent dips*

Unless a cross-section line cuts the strike of a bed at right angles, the angle of dip must be modified in the cross-section because the *apparent dip* in the plane of section will always be less than the true dip. Apparent dip can be determined by graphical methods, trigonometrical methods, or more easily, by conversion tables or charts (Badgley 1959; Berkman 1976; Compton 1962). The *Wulff* stereonet—the geologist's 'slide rule'—can also be used (Phillips 1971).

Fig. 9.2 Columnar section.

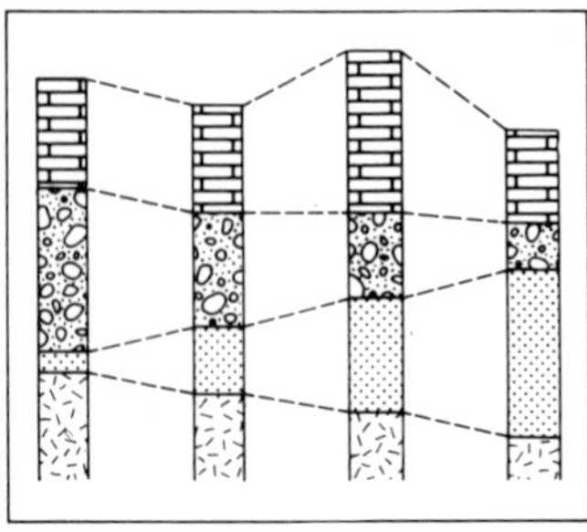

9.3 Columnar sections

Columnar sections consist of a number of simplified stratigraphical columns shown side by side to illustrate how stratigraphy or lithology changes from place to place (Fig. 9.2). They are prepared from surface outcrops and drillhole logs.

9.4 Three-dimensional illustration

Three-dimensional diagrams greatly help readers of reports to understand the solid geology described. Their preparation may help your own understanding too. There are two basic projections: *isometric* and *oblique*. Both are simple to construct.

9.4.1 *Isometric projection*

Isometric projection has no vanishing points: all parallel lines remain parallel in the diagram. The two horizontal coordinates are inclined at 30° to the E–W baseline. The viewer sees the faces of a cube, for instance, as three equal-sided parallelograms (Fig. 9.3a). Geological information drawn on the faces of an isometric block must be distorted to fit them. Grid

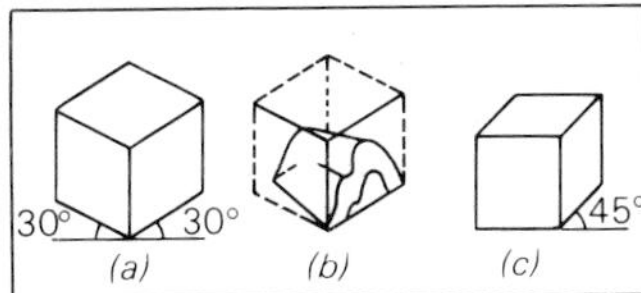

Fig. 9.3 (a) shows an *isometric* cube; (b) a fold drawn by using the cube as a guide; and (c) an *oblique* projection of the same cube.

your geological information—map or cross-section—and also the faces of the block and then transfer information by eye, grid square by grid square, or by rectangular coordinates. You can also use isometric projection as a framework for illustrations. Fig. 9.3b shows the principle, Fig. 5.15 a typical result.

9.4.2 *Oblique projection*

In oblique projection the front face of the basic cube remains a square and lies in the plane of the paper (Fig. 9.3c). The side and top faces, however, are parallelograms inclined at 45° to the E–W base-line, with distances receding from the viewer foreshortened by one-third to prevent the cube appearing rectangular. Oblique projection is ideal as a framework for showing serial cross-sections, for each section can be drawn true and undistorted (Fig. 9.4).

Fig. 9.4 Oblique projection of serial cross-sections. Each section is a true cross-section, but the distance between them is foreshortened by one-third.

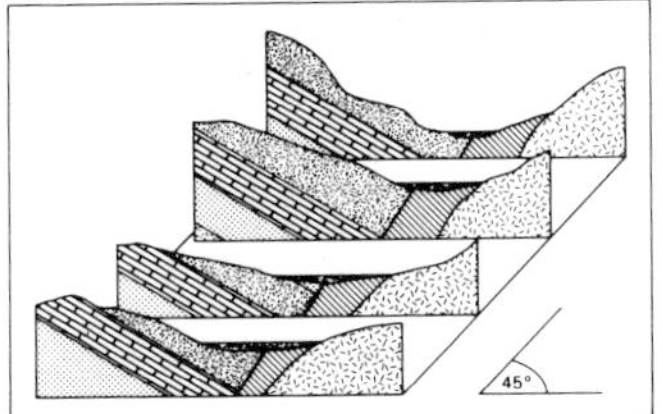

9.4.3 *Block diagrams*

Block diagrams show the solid geology of a rectangular block of ground. Either isometric or oblique projection can be used and 'cut-aways' made to bring out structural details (Fig. 9.5).

9.4.4 *Fence (panel) diagrams*

Fence diagrams—as distinct from *columnar sections*—are three-dimensional illustrations. Stratigraphy or lithology is shown on 'fences' or 'panels' connecting the different sites. Again, either isometric or oblique projection may be used (Fig. 9.6). They are often constructed from borehole logs.

9.5 Models

Geological problems can sometimes be solved by constructing three-dimensional models which can be

Fig. 9.5 Simple block diagram, split with the two halves separated, and with steps.

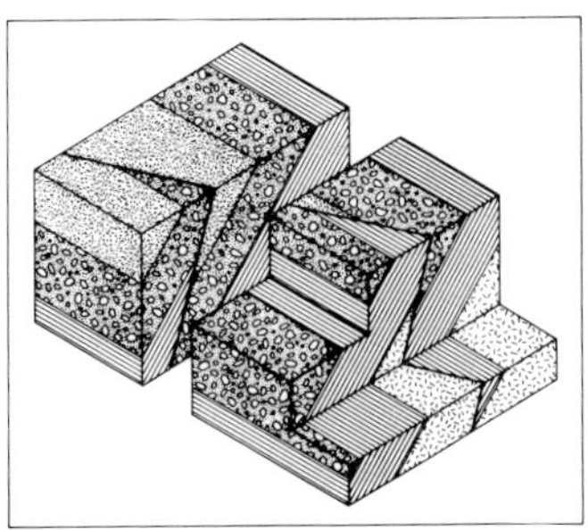

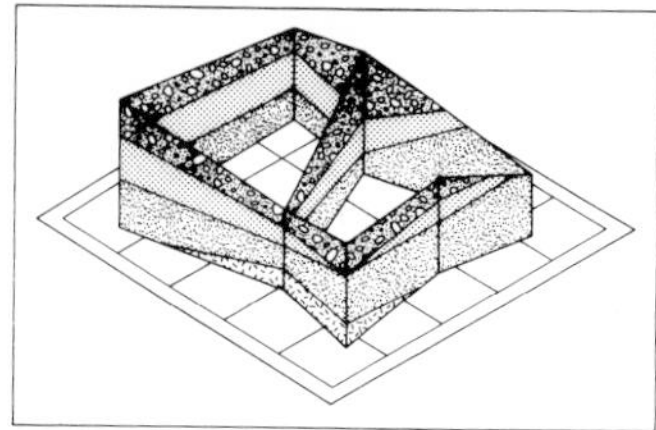

Fig. 9.6 Fence or panel diagram illustrating stratigraphic variations in a region. It gives a better spatial impression than a columnar section (see Fig. 9.2).

viewed from every direction. They need no embellishments if used only as interpretative aids.

9.5.1 *'Egg-crate' models*

An egg-crate model is constructed from a series of intersecting cross-sections. Draw the sections on bristol board or heavy cartridge paper; then cut them out and slot them together by cutting slits where they intersect. Egg-crate models are particularly useful in mountainous regions where the geology has been complicated by *nappes* and overthrusting (Fig. 9.7).

9.5.2 *Glass-sheet models*

For solving large-scale localized problems in depth models can be quickly made from window glass or perspex sheets. Trace parallel cross-sections onto glass sheets with the pens used for overhead projector transparencies. Support them vertically in a grooved, trough-like, open-ended box. Corrugated cardboard stuck to the sides of the box will usually give grooves in positions sufficiently accurate to serve the purpose of the model.

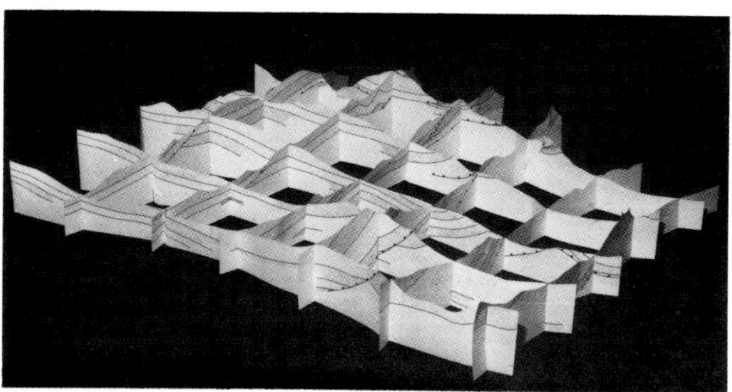

Fig. 9.7 Photograph of an 'egg-crate' model (courtesy of S.J. Matthews).

Appendix I
Safety in the field

Geological fieldwork is not without hazards. In Britain field safety is covered by the Health and Safety at Work Act, 1974. Both employers and workers have obligations under this act and they extend equally to teachers and students. A brief list of *do's* and *don'ts* for the field is given here:

1. Do not run down hills.
2. Do not climb rock faces unless it is essential to do so, and then only if you are a trained climber and you have a friend present.
3. Do not enter old mine workings or cave systems except by arrangement, and always in company. Use proper lighting, headgear and clothing and ensure that someone knows where you are, when you went underground, and when you are expected back. Report your return so that people do not arrange unnecessary search parties.
4. Always work in pairs or in close association in rugged mountains. Wear easily seen clothing.
5. Always wear a safety helmet in quarries, under steep cliffs and scree slopes, and underground, and wear goggles when hammering rocks. These are legal requirements in Britain under the Health and Safety at Work Act.
6. Never use one hammer as a chisel and hit it with another. Hammers chip: use only a properly tempered chisel.
7. Do not hammer close to other people.
8. Never pick up unexploded explosives or blasting caps from rock piles. Report them to an official if you do see them. Do not pull at pieces of fuse or wire protruding from a rock pile: they may have unexploded charges at the other end.

Whenever possible note weather forecasts in mountainous country and if you are going into a remote part of an area leave with a responsible person your route map and the time you expect to return. If mist comes down when you are in mountains, do not panic. If you are on a path and the mist is light, keep to the path. If the mist is heavy, stay where you are until it clears. The same applies if you are caught in the dark. If you are lost in mountains or on moors in *clear weather*, follow the drainage, it will usually bring you to habitation, but be careful of sudden drops on mountain streams. Forest tracks can be difficult; one looks much like another. 'Blaze' your trail if necessary, and learn to recognize your own footprints; they often help you to recognize the path you came in by.

I.1 Emergency kit

All geologists should take a course in first aid and keep a first aid kit and manual in camp with them. In very remote areas, Eastman's *First Aid Afloat* (1977) is recommended. Carry a small emergency kit in your rucksack, including dressings for blisters, a whistle and a flashlight for signalling (and a mirror if your compass does not have one). Include, also, matches sealed in a waterproof plastic bag, and an aluminized foil 'space blanket' (it weighs almost nothing). In hot climates, carry a water bottle and a packet of effervescent water sterilizing tablets. Always carry some form of emergency ration in case you have to spend a night on a hillside in mist or snow. One of the very bitter forms of 'sportsmen's' chocolate is best as it will deter you from eating it until really necessary. With sugar, it can be made into cocoa if you carry a metal cup. Take glucose tablets or similar 'pick-me-ups' with you too, to give that extra bit of energy towards the end of a long hard day.

I.2 Distress signals

The accepted field distress signal is six blasts on a whistle or six flashes with a mirror or flashlight, repeated at minute intervals. Rescuers reply with only three blasts or flashes repeated at minute intervals to prevent rescue parties homing in on each other.

I.3 Exposure

All geologists working in temperate or cold climates, and in mountains anywhere, should learn the dangers of 'exposure' (*mountain hypothermia*). It can be fatal. Exposure results from extreme chilling. It is not confined to mountains, nor is it limited to the winter months: sudden drops in temperature can occur at any time of the year on any high ground. Geologists are particularly at risk because they work in weather in which other people stay indoors. Learn to recognize the symptoms of exposure in both yourself and your companions and know how to treat it.

Prevention is largely a matter of proper field clothing. Too often, students short of money economize on equipment. This is false economy. So is the lack of a good breakfast before going into the field, or saving on the cost of food for a midday meal. Warm waterproof clothing, good boots and adequate food, all contribute to keeping warm. Do not forget a hat, for heat is lost through your scalp more quickly than from any other part of your body. It is not, however, only cold weather that causes exposure. Wind increases the effects of cold: at 0°C a wind speed of 16 km/hr produces an effective temperature of –8°C, or –14°C (7°F) at twice that speed. Wet clothing intensifies the problem, chilling by evaporation at even quite modest temperatures. Make sure your clothing is both *waterproof* and *windproof*.

Victims of exposure are not always aware of what is happening to them. If a person lags behind, resents attempts to hurry him, constantly stumbles, slurs his speech and shows a lack of interest in everything, take shelter. Get out of the wind. Get dry clothes on him if possible, if not, cover him with windproof materials, such as a space blanket. Get him into a sleeping bag or an 'emergency bivouac' if you have one: even get in with him. If possible give him a hot sugary or glucose drink but *do not give him alcohol*—it can *kill* him. Alcohol dilates the smaller blood vessels so that blood flows to the extremities more rapidly and accelerates heat loss.

If the victim is in a state of total collapse, get help quickly for if his temperature drops below 31°C (88°F) only medical treatment can save him. If you carry him on a stretcher, keep his head lower than his feet. Back at base, put him fully dressed (to reduce shock) into a bath at 45°C (113°F) for 20 minutes, *providing* his temperature is not below 31°C. If his temperature is under 31°C, get him to a hospital but, if all else fails, allow him to warm *slowly* in a warm room.

I.4 Health in warm climates

If you intend to work in a warm climate, whether tropical or not, familiarize yourself with the elementary rules of tropical hygiene and ensure that you have all the vaccinations and inoculations required by the country you are going to. If you are going to a malarial area, obtain medical advice so that you can start taking antimalarial drugs at least a week before departure. Typhoid-paratyphoid and antitetanus inoculations are highly desirable if you are to live under field conditions. Your doctor may advise inoculation against cholera and hepatitis too. Also ask him to prescribe tablets for those stomach upsets that travellers can seldom avoid. Do not rely on patent medicines, many are ineffective, some are harmful.

In camp, ensure a pure water supply by boiling or filtering your water, or both. Carry an adequate water bottle in the field and do not drink from springs and streams unless you are sure of their purity. Village wells are particularly suspect. Carry effervescent water-purifying tablets against emergencies. When travelling, drink only tea, coffee, or well-accredited bottled soft drinks: it works out cheaper in the long run. *The Preservation of Personal Health in Warm Climates* is an excellent pocket guide which can be bought from the Ross Institute of Tropical Hygiene, Keppel Street, London WC1. It covers a wide range of topics including the treatment of bites from snakes, scorpions and other pests.

I.5 Students in the field

Special considerations affect students in the field. A supervisor with a group cannot watch everyone in his party all the time—they may be scattered over a wide area. He does, however, have some responsibility for their safety and must refuse to have anyone with him who is not equipped with boots or clothing suitable for the conditions of the excursion, or who wilfully disobeys safety instructions. Otherwise, he could be deemed negligent in the case of an accident. Students may also be asked to 'sign off' at a checkpoint at the end of a day's fieldwork to ensure that no one is left behind, lost or injured, on a hillside. The checkpoint is also a convenient place to keep a comprehensive first aid kit, packed in a clearly-labelled plastic bag. It should include a mountain stretcher, an 'emergency bivouac' (a large plastic bag sold by sports shops), a flashlight and a first aid manual.

Students engaged on independent mapping must look after their own safety. There is no one to check whether they are properly clothed, wear goggles, or use their safety helmets under cliffs and scree slopes. That must be taken on trust. Even so, a supervisor still has some responsibility and may later have to justify his decision to send into the field a student who has proved incapable of looking after his own safety.

I.6 Bibliography

ANON, (undated) *Mountain Hypothermia*, British Mountaineering Council, Chameleon Press.

ANON, *Preservation of Personal Health in Warm Climates*, (1973) Ross Institute of Tropical Hygiene, London.

BROWN, T. and HUNTER, R. (1977) *The Spur Book of Survival and Rescue*, Spurbooks, Boune End.

EASTMAN, P.E. (1977) *Advanced First Aid Afloat*, Cornell Maritime Press Inc.

Appendix II
Adjustment of a closed compass traverse

A compass traverse seldom closes without error, e.g., a traverse started at A, passed through turning points B, C, D, E and F, and finished back at the starting point A again (Fig. II.1). On plotting the traverse, the last leg F–A failed to close back on A by a *closure error* (A–A′) of 110 m. To adjust, draw lines parallel to A–A′ through each of the turning points B, C, D, E and F. Distribute the error of 110 m at each turning point in proportion to the *total* distance travelled to reach that point;

Closure error A – A′	= 110 m
Total traverse distance	= 3600 m
Correction factor per traverse metre	$= \dfrac{110}{3600} = 0.03$ metres

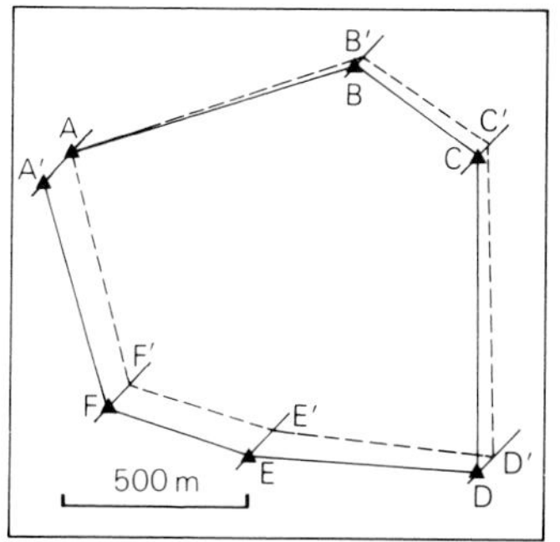

Fig. II.1 Adjustment of a closed traverse.

References and further reading

ALLUM, J.A.E. (1966) *Photogeology and Regional Mapping*, London, Pergamon.

BADGLEY, P.C. (1959) *Structural Methods for the Exploration Geologist*, New York, Harper Bros.

BATES, R.L. (1969) *The Geology of the Industrial Rocks and Minerals*, New York, Dover.

BERKMAN, D.A., RYALL, W.R. (eds.) (1976) *Field Geologist's Manual*. Monograph 9, Australian Inst. Min. Met.

BISHOP, E.E. et al. (1978) *Suggestions to Authors of the Reports of the United States Geological Survey*, (6th edn), Washington, US Government Printing Office.

COHEE, George V. (Chairman) (1962) *Stratigraphic Nomenclature in Reports of the U.S. Geological Survey*, Washington, US Government printing office.

COMPTON, Robert R. (1966) *Manual of Field Geology*, New York, Wiley.

DIETRICH, R.V. & SKINNER, B.J. (1979) *Rocks and Rock Minerals*, New York, Wiley.

FLEUTY, M.J. (1964) 'The description of folds', *Proc. Geol. Ass., Lond.* 75, 461–92.

GEOLOGICAL SOCIETY (1972) 'A concise guide to stratigraphical procedure', *Journal of the Geological Society London*, 138, 295–305.

LAHEE, F.H. (1916) *Field Geology*, New York, McGraw-Hill, (and in numerous subsequent editions).

LILLESAND, T.M. & KIEFER, R.W. (1979) *Remote Sensing and Image Interpretation*, New York, Wiley.

PHILLIPS, F.C. (1971) *The Use of Stereographic Projection in Structural Geology*. 3rd Edition. Edward Arnold (Publishers) Ltd.

RAY, R.G. (1960) *Aerial Photographs in Geologic Interpretation and Mapping*. U.S. Geol. Surv., prof. paper 37.3. Washington. U.S. Govt. printing office.

READING, H.G. (1978) 'Facies'. In *Sedimentary Environments and Facies*.ed. H.G. Reading. Oxford, Blackwell, 4–14.

REEDMAN, J.H. (1979) *Techniques in Mineral Exploration*, London, Applied Science Publishers.

ROBERTSON, R.H.S. (1961) *Mineral Use Guide: (Robertson's Spiders' Webs)*, London, Cleaver-Hume Press.

THE INSTITUTION OF GEOLOGISTS (1980) *The Geologist's Directory, 1980*. The Institution of Geologists, London.

TUCKER, M. (1982) *The Field Description of Sedimentary Rocks*, Geological Society Handbook, Open University Press, Milton Keynes.

WALLACE, Stewart R. (1975) 'The Henderson ore body—elements of discovery, reflections', *Mining Engineering* 27(6), 34–36.

亦舒精选

烈火

（加拿大）亦舒／著

南海出版公司
二〇〇二·海口

著作权合同登记号
图字：30－2002－123

图书在版编目（CIP）数据

烈火/（加）亦舒著．－海口：南海出版公司，2002.10
ISBN 7-5442-2254-3

Ⅰ．烈… Ⅱ．亦… Ⅲ．长篇小说－加拿大－现代
Ⅳ．I711.45

中国版本图书馆 CIP 数据核字（2002）第 062948 号

LIEHUO
烈　火

作　　者　（加）亦舒
责任编辑　袁杰伟　杨　雯
封面设计　姚　荣
版式设计　零语设计
出版发行　南海出版公司　电话（0898）65350227
社　　址　海口市蓝天路友利园大厦 B 座 3 楼　邮编 570203
经　　销　新华书店
印　　刷　唐山市润丰印刷有限公司印装
开　　本　850×1168 毫米　1/32
印　　张　6.625
字　　数　120 千字
版　　次　2002 年 10 月第 1 版　2002 年 10 月第 1 次印刷
印　　数　1～5000 册
书　　号　ISBN 7-5442-2254-3/I·463
定　　价　13.00 元

烈火与言诺这两个人，夏荷生先认识言诺。

而言诺与烈火之间，已存在着十多年的友谊，他俩是一起长大的。

言诺这样形容给烈火听："那样精致的脸却配那样高的身材，声音悦耳，笑容无邪，她叫我害怕；天下怎么会有那么好看的女孩子。"

说这话的时候，言诺的脸枕在手臂上，语气惆怅，眼神迷惘，像是堕入云里雾中，不能自拔。

烈火说："你恋爱了，该死。"

言诺笑笑，不置可否。

烈火惋惜地说："你应该多看看，漂亮的女孩子本市少说也有十万名。"

言诺比较内向，只说："不一样？"

"都一样。"烈火笑嘻嘻地答。

这个时候，言诺在华南刚升入三年级，荷生比他低一班，烈火在纽约大学，只有在假期才回来。

言诺常跟荷生说起他的朋友烈火。

渐渐地，荷生知道他俩的关系不比寻常。

把陆陆续续听来的细节综合在一起，荷生得到的资料是这样的：言诺的父亲是烈家的老臣子，服务超过二十年，甚得器重。言诺与烈火在小学时期已是同学，念的是本市华洋杂处的男校，英童顽皮，且已学会仗势欺人，若不是烈火处处护着言诺，只怕他吃不消要转校。

直到有一日，烈火淌鼻血青肿着眼回家，烈家才发觉校园不是安乐土。说也奇怪，家长并没有带着小孩去见老师，反而立即传功夫师傅来教咏春拳。烈火拉着言诺一齐练，小孩嘛，听见学会了可以打人，马上尽心尽意地学习，结果直到小学毕业，洋童都不敢近身。

荷生喜欢这段小插曲，烈家家长倒真有一手。

中学时期他俩一起游泳打球旅行，荷生肯定他们还齐齐考试作弊约会女孩，但这些言诺都不肯承认。

言诺笑说："我们像手足。"

荷生知道言诺没有兄弟姐妹，于是问："烈火也是独生子？"

言诺迟疑一下，"不，他有一个哥哥与一个妹妹。"

荷生一直没有机会见到烈火。

她听过他的声音，他找言诺，碰巧荷生接电话，他便活泼地说："我知道你是谁，你是诺兄夏日那枝清香的荷花。"

荷生不与他搭讪，只是笑着唤言诺来听。

荷生的母亲渐渐喜欢言诺。

“这样忠厚的人家，这样好性情的男孩子，荷生，毕业后做两年事好组织小家庭了。”

荷生与母亲一样想法，婚后生一个孩子足够，不要那种过度精灵的小大人，要笨笨胖胖的、一粒水果糖便逗得他手舞足蹈的小家伙。

她与她母亲都不知道命运另有安排。

夏荷生并没有如愿以偿。

夏荷生走的，完全是另外一条路。

那个三岔口的起点，是一个平凡的星期六下午。

言诺来接她，两人约好去看电影。

言诺一进门便兴奋地说：“荷生，烈火回来了，这次我们三个人一定要痛痛快快地聚一聚。”

荷生笑道：“久闻其名，如雷贯耳。”

“来，我们到烈府去。”

“我以为大小姐才要人接。”

言诺笑说：“我顺便替父亲送份文件上去。”

荷生当下便问：“公私能否分开？”

言诺有些遗憾地答：“怎么分？暑假我便要去烈氏企业实习，毕业后肯定进烈氏服务。”

荷生想一想，听上去一点破绽都没有。

到达烈宅，荷生一见便欢喜，只见大屋门边墙上写着一九四九“琪园”，可知是幢旧房子，荷生像时下所有的年轻人一样，怀旧心重，最爱古老事物。

玄关非常非常的深，黑白大理石地台放着一张高几，几上大水晶瓶里插满白色的鲜花，香气扑鼻。

荷生发呆，她好像来过这里，不知在什么时候，她偷偷到过这幢大宅做过客人，所以此情此景有点熟悉……

“荷生，来，到这边坐，我去找烈火。”

荷生到偏厅选一张角落里缎面子的沙发坐下。

这个地方，只有一个用途：让客人舒服地坐着等主人下来。

男孩子同男孩子到底容易做朋友，荷生没想到烈家这么富有。

换了任何一方小气些，友谊势必不能维持。

佣人放下一只茶盅，轻轻退出。

荷生刚巧戴着母亲的旧腕表，这种计时配这个地方，假如再换上一袭旧旗袍，就复古成功了。

一扇水晶玻璃嵌的长窗直通到花园去，窗门半掩，荷生忽然听到一男一女的争吵声，压得很低，却意外地清晰。

“二哥要我答允他不再见你。”

“他可以代你做主？”

“请放开我，我不想看到父亲进一步对付你。”

“父亲？父亲，嘿嘿嘿嘿。”

荷生有点不安。

她最怕类似的尴尬事，好像故意躲在角落窃听似的。

荷生马上站起来，这时玻璃门被人推开，一个女孩子匆匆跑进来，一见有人，如皇恩大赦，不管是否认识，一味往荷生身后躲。

荷生本来不是爱管事的人，不知怎地，一眼看到那女子娇怯秀美的脸，竟很自然地挡在她面前。

不出所料，有人追上来，看到偏厅内站着个正气凛然的陌生人，倒是一呆。

荷生身后的女孩趁这机会一溜烟儿似的从正门逃出去。

那个男生坐下来，细细地打量荷生。

荷生不禁恼怒，这是谁？鲁莽而无礼。

没想到对方先问："你是谁？"

荷生瞪住他。

他比言诺的年纪要大一点儿，瘦长个子，苍白的面孔，此刻正取过香烟点上。再严格的眼光都会承认他不失为一个英俊的年轻人，但不知怎地，荷生觉得他有些地方不妥。

他忽然抬起头来，轻轻吐出一口烟向荷生笑一笑。

荷生手臂上爬满鸡皮疙瘩。

这人有一双布满红丝的眼睛。

荷生反问："你又是谁？"

那人慢条斯理地答："我姓烈，你说我是谁。"

荷生吃一惊，深深失望，这便是烈火？这不是一个健康快乐的人，她不相信言诺会同这样一个人做了十多年的好朋友。

荷生脱声问："你是烈火？"

那人闻言色变，仰起头来，直视荷生，荷生被他目光中的怨怼恨恶吓一大跳，不由得退后一步。

这时候有人拉住她的手，荷生几乎喊出声来，一看，原来是言诺，这才放下心来。

只见言诺给她一个眼色，再向那人点点头，拉着荷生便走。

到了大门外，两人才不约而同松一口气。

荷生问："那就是你的好朋友？"

"不是他！"言诺跳起来。

荷生连忙说："我也猜不是，不过，他是谁，烈火又到什么地方去了？"

言诺开动车子，驶离烈宅，才轻轻说："那是烈火的大哥。"

啊。

"烈火这家伙，明明约好我，又开小差，今天恐怕看不到他了。"

荷生却十分高兴，如果适才那人是烈火，她恐怕不能爱屋及乌。

车子驶下私家路，言诺一向小心驾驶，路边却有一辆吉普车朝着他们直冲上来，言诺急忙刹车，吉普车却不肯停，荷生大叫出来，吉普车的保险杠碰到他们的车子才不动了，荷生觉得全身血液统统涌上脑袋，吉普车司机却哈哈大笑起来，还拍着手。

荷生破口骂：“疯子！”

谁知言诺也相继大笑起来，打开车门跳出去，“可不就是烈疯子。”得意洋洋，引以为荣。

言诺与吉普车司机拥抱。

到这个时候，荷生已经知道这人是谁，她左边太阳穴隐隐作痛，烈家兄弟恐怕都有点异常。奇怪，言诺的坐言起行最平凡稳健不过，怎么会交上这样的朋友。

只见他俩嘻嘻哈哈互相拍打一番，揽肩搭背地向荷生走来。

只听得那人笑着说：“我知道了，你一定是夏荷生。”

荷生看清楚他的面孔，吓一跳，连忙侧过头去。

“荷生，这才是烈火。”

烈火与荷生打一个照脸，也是一呆，言诺这愣小子太不会形容了，单凭他的言语，也太委屈夏荷生了。

当下他摸一摸胡子，“都怕这个，难怪父亲叫我剃掉它。”

言诺伸手搓一搓他蜷曲的长发，“可以梳辫子了，穴居人似的，吓坏人。”

烈火笑着说：“荷生，你来说句公道话。”

荷生看着他，“不是叫你烈疯子吗?”

烈火叉着腰笑。荷生觉得他自顶至踵，外形上没有些微缺点瑕疵，性格活泼热情，难怪言诺说过，烈火在家中最受钟爱。

忙着谈笑，三人竟没有发觉一辆黑色大轿车已静静停下，老司机下车笑道：“烈先生请你们回屋里详谈。”

荷生好奇地回头看。

大车子深色玻璃里隐隐有个人影，想必是烈家老爷了，没想到一天之内就见齐烈家的人。

一双炯炯有神的眼睛，在车内也正在打量夏荷生。

言诺与烈火各自把车驶开让路。

老司机笑着同老板说：“年轻人，不怕热。”

荷生的白棉衫为汗所湿，胸前一搭V字汗印，额前碎发统统黏在额角与颈后。她站在一边，看大车离去。

怎么不怕热? 烈火目光中那点炽热就叫她害怕。

烈火笑着说：“来，我们回屋里谈。”

言诺说：“你问荷生可要看电影。”

烈火正要开口，一辆银灰色高性能跑车俯冲下来，言诺连忙把荷生拉至怀中，双双避开。

荷生看着车尾，惊骇地说：“这条私家路怎么这么繁忙，难怪说马路如虎口。”她轻轻推开言诺。

烈火说：“那一定是烈风。”

言诺点头：“我刚才见到他。”

烈火问：“他为何而来？”

“我不方便问。”

荷生听在耳中，不用解释也知道烈家弟兄并不和睦。

“屋里还有谁？”

“烈云在家。”

“叫烈云陪荷生去看电影好了，我有事同你研究。”

到这个时候，荷生“啊哈”一声，开了腔，“烈火你听着，言诺是我的男朋友，我不怂恿他撇下你你已经够幸运，你甭想丢下我与他单独行动，我把话先说明白了以后大家好做，我不用你替我安排节目，言诺，我们照原定计划。”

言诺大笑，“烈火，听见没有，改天见啦。”他朝好友摆摆手。

荷生马上仰起头，向烈火做出一个胜利的样子，挽着言诺的手上车。

烈火为之气结，伸脚踢言诺的车子，力道甚大，车身一震。

荷生伸出头去，“长毛！”她笑着骂他。

烈火被无数女孩子骂过千万次，早就老皮老肉，可是夏荷生这两个字，夹着银铃似的笑声，却在他耳畔萦绕，久久不散。

荷生松一口气，同言诺说："吁，那一家人。"

"你说得对，家父讲过，烈家的人，有一股奇异的魅力，一旦与他们接触，身不由己地受到吸引，真心想同他们交往。"

荷生不予置评。

言诺说："你会喜欢烈火的。"

"啊，我并非不喜欢他。"

"你要把他当兄弟看待。"

"你俩真的那么要好？"

"真的。"

荷生摇下了车窗，任由热风吹进来。

一直到电影散场，她都没有说过什么。

那天晚上，她躺在自己的小卧室里。半夜，仿佛有热风吹拂脸庞，又像有一个人，不停地用手抚摸她的鬓角。荷生迷惘地抬起头来，看不清楚他是谁，但她肯定那不会是言诺，那手强壮而温暖，荷生没有拒绝。

天蒙蒙亮时她醒来，靠在小床上，呆了半晌。

她拨电话给言诺。

言诺还没有醒，听到女朋友的声音，很愉快地说："我做梦看到你。"

奇怪，荷生想，她的梦境里，从来没有言诺。

"今天我到烈家去，你要不要一起来？"

"不，"荷生说得很坚决，"你们玩好了。"

"我介绍烈云给你认识。"

“不要，我不寂寞。”

“荷生，你好像很抗拒烈家的人。”

是的，荷生觉得他们危险，同性格这样强的人，要维持一个安全距离，才能避开压力。

“烈云跟你们一起？”

言诺误会了，他笑着解释，“她刚学走路时我已经认识她，荷生，你不必多心。”

稍后荷生站在露台上，用手摸一摸心房，心不怕多，只怕它偏，切莫偏到腋下去才好。

言诺忘记这是荷生二十一岁生日。

去年认识小言的时候也是六月六日，图书馆里，他的书包同她的调错了，他比她早发觉，因阻迟他替小孩子补习的时间，非常气恼地追上来，一把搭住荷生的肩膀，大声吆喝着“喂你等等”，荷生皱着眉头转头去，说也奇怪，小言的火气顿时消失，眼目如被贴上清凉剂，呆半晌，他说：“还你书包。”

这傻小子结果没去为任何人补课，他一直跟在荷生身后，亦步亦趋，她上公路车他也上，她下他也下，结果荷生打圆场：“你是二年级的言诺吧？”他功课十分好，在校园小有名气，这次派上用场。

他们到附近的饮冰室去坐下，他请她吃红豆冰，而在稍后小言才懂得庆幸荷生不是动辄要坐大酒店咖啡厅的女孩子。

足足一年了，若没有几个考试支撑着，更不晓得日子是怎么过去的。

自从小言在她身边，好几个科目的成绩都突飞猛进，她相当倚赖他，每天通两次电话，芝麻绿豆都报告一番，少女有时会为很小的事情生气，小言有说不出的诧异，总是劝道："不要在乎别人说些什么。"他的口头禅是"管它呢"。

就在上个月，小言把她带回去见过父母。

一进门就知道是一户正经好人家，自置公寓打理得一尘不染，有一位老佣人管小言叫大官，做得一手好粤菜。

言太太是位爱打牌不理事的中年妇女，不讲话也有点眉开眼笑的样子。

她同荷生说："我们只有他一个孩子，年前房产跌价时他父亲在山腰置了层小公寓，预备他成家用。年轻人都喜欢住那一带，最要紧是清静，交通不便也没关系。"

荷生听得懂。

那是告诉他们，随对可以注册。

出来的时候小言说："他们喜欢你喜欢得不得了。"

荷生很愉快。

夏太太更加高兴。

荷生在星期天一向有赖床的习惯。

夏太太探头进房，"荷生，言诺找你，他说二十分钟后来接你。"

“不！”荷生脱口叫出来。

夏太太莫名其妙，“吵架了？”她坐在女儿床沿，“这样好的男孩子……你要爱惜他。”

荷生微笑，“你一直帮他。”

“因为他一直帮你呀。”

荷生到浴室掬起冷水洗脸。

“待你婚后我就到加拿大去跟你姨妈生活。”

“你现在就可以去，我早就能够照顾自己了。”

“唉，其实我是舍不得这个城市。”

男女老幼都不舍得，已经不是新闻。

忽然之间，楼下汽车喇叭声大作。

“这是哪一家的阿飞？”夏太太探头出窗。

夏家住老房子，没有几户人家，只见好几个屋主都在张望。

荷生心中有数。

夏太太讶异地说：“荷生，你快来看，是小言同一个阿胡髭在一起，这是怎么一回事？”

荷生笑，“所以呵母亲，别把事情看得太简单，言诺这小子也有另外一个面孔。”

“那野人是谁，一脸的毛不怕长痱子？”

荷生预备出门。

“小言怎么会有那样的朋友，近墨者黑，近朱者赤，要小心谨慎。”

荷生开门，“早就来不及了。”笑着关上大门。

在梯间她听到喇叭声震天地响。

荷生对言诺说:“你太纵容他了。”

言诺笑,“这是他庆祝你生日的前奏曲。”

烈火自司机位探头出来,“夏荷生,自今日开始,你已是个成年人了。”

荷生避开他的目光,“小言,你来开车。”

言诺与烈火换了位子。

“荷生,今天由烈火替你安排节目。”

荷生冷冷地说:“我的生日由我自己安排。”

烈火咯咯地笑她反应过激。

言诺也笑道:“我们一整天都喝香槟,不用其他饮料。”

都是那个人的馊主意,荷生有种感觉,她与小言之间所有的宁静会叫他给破坏掉。

荷生坐在前座,老觉得脖子后面痒丝丝,似有人在她颈后哈气,她忍无可忍,别转头,正待苛责烈火,却发觉他舒舒服服躺在坐位里,用一顶破草帽遮住脸,正在假寐。

不是他。

荷生疑心自己生了错觉。

她涨红面孔,连耳朵都麻辣辣地烧起来,颈后那只无形的手竟大胆地伸过来拨弄她的鬓角,她拂之不去,浑身起了鸡皮疙瘩。

是同一只手,昨夜那只手。

荷生低下头,闭上眼睛,原来真是真的,已经来不及了。

车子停下来。

荷生张开眼睛。

她听见后座有人懒洋洋地问：“烈云出来没有?”

言诺笑说：“在玻璃门里边，她看到我们了。”

荷生朝他指向的方向看过去，想像中的烈云是个卡门那样的女孩子，同她兄弟差不多脾气，但是玻璃门内穿白衣的身形十分熟悉。

她推开车门，“我去叫她。”

烈云正与几个同龄女友说话，她们一个个打扮得花团锦簇。

走近了，荷生发觉烈云竟是那日在烈宅偏厅躲在她身后的女孩子，不禁又添一分惊讶。

这个时候的烈云，却是另外一副面孔，细软的短发全部似男孩子那样梳往脑后，一袭露背白裙，闻声转过头来，看到荷生，她也认出了她。

那群女孩子忽然一阵骚动，原来两位男生也跟了进来，她们朝异性迎上去。

荷生十分感喟，条件那么好的女孩子，本市女性人口比男性又是一比一点二，何用这样心急。

她听得烈云说：“我知道你是夏荷生，言哥哥的女朋友。”

语气天真而清脆，夏荷生马上喜欢她，亲切地说：“那么叫我夏姐姐。”

烈云只是笑。

烈火先走过来，“我们在这里订了位子，烈云，参加我们好不好?”

“我只有时间喝一杯香槟。”

荷生看一看那边，言诺让女孩子围得紧紧的。

烈火说：“我们先过去坐下，烈云，你等言诺。”

荷生走在他身后，离一截路。这是间私人会所，装修非常考究，地板是柚木格子，偏偏烈火与荷生两人都穿着球鞋，没有半丝声响。

走廊很长，走着走着，烈火起了疑心：她还在身后吗？终于忍不住，回头看荷生。

荷生见他停步，也就站在原地。

两人静静对望一会儿。

烈火说：“奥菲斯不该往回看。”

荷生答：“别担心，幼罗底斯不在此地。”

言诺走上来，笑道：“你们俩肯说话就好。”

不知怎地，烈火与荷生异口同声地说：“我们一直有说有笑。”

言诺笑，“哦，是吗?”

荷生别转头去。

烈云说：“父亲应允我，当我二十一岁的时候，给我一间公寓，让我搬出去住。”

荷生喝一大口香槟定神。

烈云放下杯子，“她们在等我呢。”

言诺站起来送她，“玩得开心点。”

烈云跟荷生说："生日快乐。"

荷生连忙答："谢谢你。"

烈云一转身，裙裾扬开，色如春晓。

荷生赞叹，"令妹是位美女。"

烈火笑，"我是野兽。"

这并不正确。

他们一家都长得美。

如果你相信优生学的话，便可以肯定烈老爷栽培这样漂亮聪明的孩子是特别用过心思的。

但敏感的荷生始终觉察到他们三兄妹似乎有许许多多难言之隐，明媚的表面底下不知收藏着什么样的黑暗危机。

她想得太多。

平静朴素的大学生活忽然闯进烈火这样一个人，使得荷生遐思不断。

"荷生，荷生。"言诺叫她。

荷生拉一拉衬衫领口，"这热浪叫我疲倦。"

言诺笑，"热？空调畅通，何热之有？"

烈火却说："用力抗拒一件事的时候，最消耗精力，一下子就累得贼死。"

荷生问自己：你在努力抗拒什么？夏荷生，说呀，你是知道的，你只是不敢说，你只是不敢承认。

言诺说："烈火，你的话最多，快介绍一下自己。"

“我？我是言诺的好友。”

糟糕，夏荷生想：我是言诺的女友。

“奇怪，”言诺取笑他，“平日你滔滔不绝，宏论最多，绝无冷场，今日水准大跌，令人失望。”

烈火并无自辩。

言诺勤于制造话题：“把你经营的花圃告诉荷生。”

荷生抬起眼睛，这倒是一个风雅的嗜好。

言诺说：“我来讲也一样，不对的时候你更正我。”

烈火笑，“少年时的玩意儿，好久没去花工夫了。”

荷生好奇，一定是个玫瑰花圃，现身说法。

“来，烈火，带我们参观一下。”

花圃在烈家后园一角。

小言说：“要不是父亲逼着他去念商管，或许烈火会成为植物学专家。”

在言诺眼中，烈火没有缺点。

车子驶抵烈府，这是荷生第二次来了。大宅静悄悄，烈火带着他们自后门走，花园对着山下蔚蓝的大海，面积比荷生想像中的大。

她没有看见嫣红姹紫的花圃。

荷生不动声色，她知道言诺与烈火在考她。

她慢慢走到石栏杆处靠着看风景。

她听到烈火轻笑。

荷生低下头，发觉左脚踩住一棵草本植物，茎是方形的，卵形叶子对生，被她踏碎部分发出一股清凉的香气，荷生低头摘一片叶子揉碎嗅一嗅，觉得沁人心脾，顿时凉快不少，她诧异地说："薄荷。"

烈火笑，"被你猜中了。"

荷生大感兴趣，"其余的是什么？"

看仔细了，她发觉有许多种植物，大半拥有貌不惊人的小叶子，言诺一一为她数出来：甘草、天麻、地黄、黄连、艾、菖蒲、茯苓……

荷生高兴到极点，蹲下来逐一细赏——"艾叶与菖蒲在端午节燃烧薰屋，传说可驱邪除病"，"甘草是中药中的百搭呢。"

她忽然看到角落有两株绿茎碧叶的白花，花形像牵牛花，但是比牵牛花大，于是问："这是什么？"

烈火答："曼陀罗花。"

"什么，这就是鼎鼎大名的地狱之花？"她后退一步。

言诺说："它也是一种药用植物。"

烈火笑，"是一只为盛名所累的麻醉剂。"

荷生惊叹，"我可以在此地研究一整天呢。"

佣人捧出冷饮，烈火与言诺走到泳池旁的太阳伞下面。

荷生抬高头，正在欣赏一边墙上爬得满满的常春藤，忽然之间，她后颈那股麻痒的感觉又来了。

荷生吓一跳，猛地转身，一边用手去拂扫，却看到二楼露台长窗内站着一个人。

那人随着荷生的目光隐失。

荷生惊疑地搓揉着后颈。

小言在那边招呼她，“过来歇一歇，当心晒坏了。”

荷生坐下喝一口冰茶。

刚才谁在窥望？

她听到烈火说：“父亲的意思是叫我留下来，明年待你毕业，我俩全力出击。”

言诺说：“我这边一点问题都没有。”

烈火答：“祝我们前途似锦。”

荷生忽然问：“烈风今天在不在？”

言诺连忙说：“不，他不住这里，他另外有个家。”宛如烈家的发言人。

荷生实在忍不住，“那么，”她伸手指一指，“谁在那个露台上？”

言诺看一看，“有人吗？”

荷生笑，“你让烈火回答我好不好？”

烈火却已经走进屋内。

言诺按住女友的手，“荷生，他们家事比较复杂，我们不便问及。”

“对不起。”

言诺想一想，还是透露消息，“他有两位母

亲。”

啊，多了跟少了都不美，荷生缺父亲，烈火多出一个母亲，只有言诺最幸福，一父一母，恰恰好。

看样子他们两兄弟同父异母。

言诺又说：“只有烈云是他亲妹妹。”

荷生见言诺代烈火遗憾不已，便安慰他说：“这样的家庭也常见。”

“烈火不这样想，不是他父亲追他，他才不肯回来。”

荷生说：“我们也该告辞了。”

言诺点点头。

刚走近长窗，就听到重物堕地声。

言诺像是知道发生了什么事，立即冲入书房，荷生跟着进去，发觉言诺已紧紧抱住烈火，不让他动弹。室内凌乱，一张红木茶几掀翻在地，另一头站着烈风，手中抓着一只椅子当武器。

烈火狠狠地说：“你给我滚出去，以后不准你进这间屋子来。”

烈风冷冷地答：“笑话，这间屋子是我外公的物业，跟我母亲的名字叫琪园，我不把你轰出去已经很好，明明是你霸占我的产业，你倒恶人先告状。”

言诺劝道：“一人少一句吧。”

谁知烈风指着他骂：“姓言的，你父亲忘恩负义，枉费周家栽培他，到头来倒戈相向，有老奴才就有小奴才，这里容不下你说话。”

言诺脸上变色。

烈火大力挣扎，“你还不松手让我赶走他。”

荷生站在一角急得好比热锅上的蚂蚁。

烈火额上青筋绽现，“你听着，再被我见到你缠着烈云，我发誓杀掉你。”

荷生不顾一切，走向前对烈风说：“请你先避一避。”声音里充满恳求。

烈风先是瞪着荷生，不知怎地，用力扔下椅子，转头走开。

言诺松开烈火。

烈火还要追上去，被荷生挡在门口，无论如何不给他过关，烈火这才倒在沙发上，不言不语。

荷生过去蹲下劝他，“喊打喊杀有什么好？像我们，想要有个同胞手足还不能够，你俩却互相作践。”她管这桩闲事，像是管定了。

烈火用手捂着脸，荷生有荷生的牛脾气，硬是要扯下他的手，言诺在一旁急得要命，他怕烈火怒气冲天，一句话或是一个动作得罪了荷生，以后无法弥补。

但是没有，烈火渐渐静下来。

室内三个人，都紧张得叫汗湿透了衣裳。

佣人到此时才敢探头进来查看。

荷生同言诺说：“我们走吧，让他休息。”

离开大宅的时候，荷生没有法子不再留意门旁“琪园”两个字。

她看言诺一眼，这里边的故事，小言当然是清楚的。言家与烈家的交情恐怕不止二十年，烈老爷怎么起的家，同两位妻子的纠葛，言诺统统知道，不过他不说，荷生不会去问他。

烈家的司机赶出来，“言少爷，烈先生叫我送你们。”

荷生讶异，“烈先生在家？”

“是，他还说，谢谢夏小姐调解纷争。”

家主在家！

他为什么不露面？

身为家长，应当出来压住场面。

言诺拉拉荷生的衣角，低声说：“烈风与烈火无日不吵。”

两人上了车。

小言又说：“两兄弟的心病不止一朝一夕了。”

荷生有一点点同情烈风，但眼见言诺与烈火站同一阵线，不便发言。

半晌她问：“园子里有没有金鸡纳树？”

外表粗犷的烈火竟会有心思经营一个中药植物花圃，真是不可思议。

到了家，小言没有送荷生上去，她另外有约，一班女同学要与她出去跳舞。

小言叮嘱她，“看到英俊小生，不得目不转睛，不可与他说话，不许与他跳舞。”

荷生问：“那么，能不能与他私奔?”

小言睁大眼睛，做一个狰狞的表情。

荷生笑着逃上楼去。

她心里却有点凄惶，到了家门，把身子靠在墙上发呆，出去几个钟头，像打过一场仗，累得说不出话来。

隔一会儿才掏出钥匙开门进屋。

躺在藤榻上就睡着了。

人影，有一个人影，轻轻地走过来，“荷生，荷生，随我来，荷生，我召唤你，随我来。”

荷生惊呼：谁，谁?

“荷生，荷生。”

是她母亲推她，睁开眼，天色已暗。

明明已睡了好些时间，为何恍似一刹那?

“同学打过电话来催，叫你准时到。”

荷生点点头。

夏太太喃喃道：“真热，汗出如浆，让我关上门窗开冷气。”

荷生坐起来，藤榻上浅浅一个汗印。

荷生怕她整个人会热融掉化成汗水。

天空霍霍闪起电光，雷声隆隆，刮来一阵雷雨风，夏太太忙到露台收衣服。

大雨忽然倾盆倒下，哗啦哗啦，四周只有雨声。

夏太太问："有没有人来接你？肯定不能穿白皮鞋了。"

荷生站在露台边，抱着手看景，一片白濛濛，气温顿时下降，凝得一屋雾气。她拂一拂脸上的水珠，回到房内淋浴装扮。

珠灰色晚服是早就预备下的，荷生来不及吹干头发就套上裙子。

雨没有停，荷生也没有期望它停下来。

穿上镂空鞋，她走到门口，回头一望，发觉她母亲在卧室内看电视，荷生微微一笑，下楼去赴约。

大雨中车子与街灯都只是一团光，荷生根本不知道她怎么样才能抵达目的地，可是也不在乎。

她站在檐篷下，原来手上连雨具都没有。

"好大的雨。"身后有人说。

荷生抬起头，并没有惊讶，宛如她一早知道他会来，她似在等他。

脚背早被雨溅湿，她没有退后。

荷生看向雨中，他站得离她很近，手臂与手臂之间像是没有缝隙，但又好像隔着一线天，荷生动都不敢动，也不能动，她已被点了穴道。

脖子后边那股酥麻的感觉又来了，这次，她肯定是他在哈气。

前有水，后有火，荷生不知如何抉择。

过了很久很久，荷生听到他说："我会同言诺讲。"

荷生落下泪来。

"我不知道会发生这样的事，我很抱歉。"

天空上雷电交加，传说人若做了亏心事，天雷会转弯搜他出来打。

荷生幼时怕行雷闪电，此刻她觉得最可怕的是她旁边那个人，不不不，最可怕的是她自己夏荷生。

他像是已说完要说的话，缓缓转身，走进雨中，双手插在衣袋里，调过头来，看着荷生，笑一笑，消失在雨里。

荷生独自站着落泪。

不知过多久，才有一辆计程车驶来，有人下了车，荷生才上去。

晚饭已吃到第三道菜，女主角方姗姗来迟，女同学起哄。

荷生嗫嚅地解释，"大雨天叫不到车。"

"小言不是你的司机吗？这回子又找什么借口。"

有人替她叫了一杯白兰地来挡挡湿气。

"生日快乐。"大家举杯。

还是同一日？荷生不能置信，感觉上像已经过了一年，两年，甚至十年，她对时间有点混淆。

有人替她拨一拨湿头发，荷生如惊弓之鸟般弹起来，恐惧地看住那只手。

女同学笑着问："怎么了？"

她连忙呷一口酒。

靠着酒力，渐渐松弛下来。

"荷生，有什么愿望？"

愿望，呵是，愿望，荷生用手撑着头，"我要三个愿望。"

"荷生，别太贪婪。"

"算了，一个人有多少二十一岁，一下子就老了，让她去。"

荷生苦苦地笑。

坐首席的女孩子一阵骚动。

"荷生，那边有位先生送香槟给我们喝。"

"呵，一定是夏荷生的神秘仰慕者。"

夏荷生已经有三分酒意，转过头去，远处一张桌子上，坐着个熟人，见荷生看他，颔首示意。

荷生吓一跳。

是烈风。

荷生连忙注意他的女伴，那女孩子穿得很暴露，正翘着嘴不高兴。荷生见不是烈云，放下一颗心。

为什么担心是烈云？好没有来由，荷生觉得她似走入迷宫，无数出路，统统是死胡同，只有一扇活门，但伪装得和其他通道一模一样。

同学问："他是谁？"

"朋友。"

“咄，一朋三千里，老老实实，我就没有请喝克鲁格香槟的朋友。”

大家一阵哄笑。

荷生再回头的时候，烈风与他的女伴已经离去。荷生发呆，他们那一家人，爱来就来，爱去就去，专门在普通人的生活中留下不可磨灭的涟漪，一圈一圈荡漾开来，到最后，凝固了，变成年轮，他们却当是等闲事。

女孩子到底是女孩子，下半场不胜酒力，局散各自回家。

荷生并没有醉，只是累。

一如所有喝醉的人，不肯承认醉酒。

一如所有怀才不遇的人，只是混赖社会。

第二天醒来，荷生先是想：哟，怎么撑得到学校去，随即觉得头痛欲裂，这才想起，她正在放暑假，可以自由地在床上再躺个大半天，于是大声呻吟。

昨天的事，一个个归队，在她思维中出现。

不住地揉着太阳穴，荷生苍白地起床找水喝。

走过客厅，看到有人端正地坐在沙发上等她。

是言诺，他没有叫她，静静地以他明亮的眼睛看着她，

荷生从来未曾笑得这么假过，“我母亲呢?”

“伯母有事出去了。”

“怎么不叫醒我?”

“推过你叫过你，你沉睡不觉。”

言诺一脸困惑，荷生当然晓得他为什么而来，她用手捂着脸，叹息一声。

“荷生，昨夜烈火来找我。”

荷生说：“我先去漱口。”

“不，你坐下来，听我把话说完。”

荷生低着头，“请讲。”

言诺应该得到一个解释。

“烈火所说，都是真的？”

荷生觉得越拖得久越是残忍，于是她鼓起勇气答：“是。”自己都觉得这个字像一把利刃，直剜入人心。

言诺要过一会儿才能说：“荷生，这是不可能的事，你认识他，还未超过一百个小时。”

荷生轻轻地说：“这不是时间上的问题。”

“你根本不清楚他的底细。”

“并不需要知道。”

“你认为你做得正确？”

“同对与错一点关系都没有。”

“荷生，我不明白。”

荷生迷惘地抬起头来，“我也不明白。”

“我竟一点不察觉，我像个盲人。”

“你责备我吧，言诺，我伤害了你。”

“这是你的错吗？未必。”

言诺的声音里混杂了悲哀、无奈、失望，但荷生听不到任何恨意，他是一个君子，在任何情况之下，他都维持着应有的风度。

言诺别转面孔，“我没有争的习惯。”

荷生答：“也不值得那么做。”

言诺站起来，不知怎地，脚步踉跄，撞向茶几，荷生欲伸手去扶他，他闪避，荷生看到他右手指节处表紫红肿。

言诺轻轻说：“我也没有打架的习惯，出气对象只是一堵墙。”

荷生声音颤抖，“相信我，言诺，这是我的损失。”

言诺微笑，“是吗，那我得到的又是什么？”

荷生不能回答，怔怔地看住他。

三个人坐一起玩一局游戏，有人赢，就一定有人输，一桌上的筹码不会不翼而飞，必然有人失有人得，怎么可能三个人一起输。

但荷生明明没有赢的感觉。

言诺伸出手来，轻轻抚摸荷生的鬓角，过了一刻，一声不响地离去。

荷生对着电视机，下午时分，正在播放动画片，猫与鼠正做永恒的追逐。荷生觉得生活中没有更悲惨的故事了，她泪流满面。

荷生为自己而哭，她不担心言诺，像他那般人才，哪里找不到贤慧美丽的伴侣。

夏日余下的日子，荷生蜷缩在屋内，不肯外

出。

她母亲完全知道发生了什么事，爱莫能助，因此不发一言。

待荷生有勇气面对现实的时候，立秋已过。

这段时间内，她不想见任何人。

返学校办新学期入学手续那日，她生怕碰到言诺，任何男生走过身边，她都吓一跳。荷生凄凉地想，惟一问心无愧的是，她不是个一脚踏两船的女人。

办完正经事，约好同学买书，走到门口，听到汽车喇叭响两下，荷生没有留意，同学转过头去看，随即笑说："找你呢，夏荷生。"

荷生吞一口涎沫，这样的作风，像煞了一个人。

她僵硬地转过头去，看到车里的人，怔住，不禁问："烈云？"

烈云探出头来，笑道："请上车来。"

烈家作风是不会变的，假设其余人等一概听他们命令，以他们为重。

荷生正犹疑，同学已经谅解地说："找到书会替你多买一本。"

荷生好生感激，那边烈云又按两声喇叭。

荷生过去上车。

"烈小姐，或许你应考虑预约。"

烈云笑道："夏小姐，这两个月来你根本不接电话。”

烈云所说属实，荷生不做声。

“别叫我下请帖，我二哥说，你不收信件。”

荷生只是赔笑。

“你瘦了许多。”

荷生转头问：“谁叫你来的？”

烈云正在拐弯，一脚踩着刹车，轮胎与柏油路摩擦，吱吱发响，她笑问：“我自己不能来？”

荷生不出声。

“好好好，是二哥让我来看你心情如何。”

这样说，他们三个人的事，烈云全知道，唉，也许烈家上下全知道了，荷生尴尬得涨红面孔。

她蓦然抬起头来，“我们到哪里去？”

“到琪园好不好？”

“不，不。”

烈云笑，“看你吓得魂不附体。”

荷生气急败坏，“烈云，这么多人当中，你首先不应开我的玩笑。”

烈云温柔地说：“你讲得对，荷生，我们第一次见面你就帮了我一个大忙。”

荷生吁出一口气。

“这样吧，我们到会所喝杯茶？”

“你送我回家岂非更好。”

烈云十分诧异，“你想躲到几时？他们已经没事了，言哥哥每天下午四时到七时都与二哥一起办公，过些时候还要一齐出差到东京去。”

荷生大吃一惊。

她像那种二次大战孤身流落丛林四十年的日军，猛地听见战争结束世界和平，不能置信，拒绝返回文明。

“让你静这么多个月还是我的主意，二哥每晚开车到你家楼下你可知道?”

荷生像在听别人的故事，不，她不知道。

没想到有人会这样做。

“有一两次我与他在一起，静静地喝杯咖啡，偶尔聊几句，并不闷。”

兄妹俩坐在车子里，借月色仰视露台，盼望伊人出现，可惜的是，有露台的人家通常不到露台站，只有佣人淋完花之后晾衣服。

烈云问哥哥：“她出来你打算怎么样?”

烈火把头伏在驾驶盘上，“我不晓得，或许吹口哨。”

烈云骇笑，“可是你从来没有怕过女孩子。”

烈火口气也带绝望的意味：“我知道，这次我在劫难逃。”

烈火从来没有这般严肃过。

“他熬过一段相当痛苦的日子。”烈云说。

到了俱乐部，烈云去停车，荷生先找到台子坐下，还没叫饮料，已经有人走过来，挡在她面前。

荷生苦笑，烈云当然是有阴谋的。

她抬起头，意外地看到白衣白裤的烈风。

“不介意我坐一会儿吧?”

“当然不。”

烈风手上拿着一杯啤酒，他彬彬有礼，为荷生叫了饮品，向她举杯，“谢谢你。”他说。

荷生奇问：“为什么?”

烈风苦涩地答：“你不歧视我，你没戴有色眼镜看我。”

“我相信其他人也没有。”

烈风说：“那边坐着的是我母亲，你能说服她便是帮我一个大忙。”

荷生斜斜看过去，只见一位身形纤弱穿着香奈儿套装的中年妇女正与朋友聊天。

呵，这便是琪园原来的女主人。

“自幼她便教我打垮那边的人。”

荷生忍不住带一点揶揄：“有没有成功?”

“我太讨厌烈火，太喜欢烈云，全盘失败。”

荷生不习惯烈家兄妹一贯惊人的坦率，觉得十分震荡，顿时有点不自在。

“我知道你今天为何而来。”

荷生一怔。

烈风笑笑，“我是闻风来看热闹的，听说你们即将举行群英会。”

荷生听懂了，知道烈火与言诺稍后会出现，不禁对烈云这个安排生气，她轻责烈风：“假使你当我是朋友，又为何挖苦我？”

烈风一怔，承认道：“你讲得对，但是我讨厌他们两人，你先后两个选择都不明智。”

荷生啼笑皆非，别转面孔，不去睬他。

“或许我在白天不该喝这么多，又可能我喝得太少，否则也可以加入战团，岂非更加热闹。”说罢嘿嘿笑起来。

荷生忍无可忍，“烈风，你语无伦次，很难怪人不喜欢你。”

他沉默下来，过一会儿拿着他的杯子离去。

荷生叹一口气，看样子烈云扔下她不打算回来了，荷生拾过书本手袋，刚站起来，烈火已经出现。

荷生觉得她似面对执行死刑的长枪队。

过半晌她问：“言诺呢？”

“他不肯来，他说他不是好演员。”

荷生反而放下心来，三个人若无其事地坐在一张桌上谈笑，未免太过滑稽，烈云的估计略有偏差，他们还未能达到这种境地。

烈火微笑，“好吗？”

荷生一时答不上来，感慨万千。

烈火的头发与胡子都修理过，外形更加潇洒，猛眼看与烈风有三分相似。

荷生转过头看，烈风与他的母亲已经离去。

烈火这样聪明的人，当然知道荷生的目光是找谁，他说："以后不要同这个人说话。"

荷生微笑，"你太习惯干涉他人的言行举止。"

"你别误会，那个人是例外。"

"他是你兄弟。"

"他尽得母系恶劣遗传。"

荷生心情本来不好，但听到如此偏见的评语，不禁莞尔。

烈火说："你笑了。"

荷生伸手摸自己的脸颊，真的，隐没多日的笑脸，终于再度浮现，可见最难的已经过去。

像传说中那种没有良知的女人，她自辜负他人的过程中获得更好的选择。

烈火说："烈云不看好我们。"

荷生不出声。

"她觉得我俩性格太强，易起冲突。"

"你跟妹妹十分亲厚，无话不说。"荷生羡慕。

"我与你之间也是一样，你可以把所有心事告诉我。"

不可能，荷生不住地摇头，他们一开头已经得不到诸神的庇佑，她伤害了言诺，内心不安，也许，连这一点点内疚都会渐渐埋没，但不是今天明天后天，目前荷生尚不能释然，光是这一点，已经不能告诉烈火。

他们走到停车场。

烈火指指一辆黑色大车，"这是烈风母亲的座驾。”

仍然维持着豪华的排场，可见烈先生在经济上照样看顾她，抑或，荷生忽然想起来，这是烈风外祖父的余荫？

多么复杂的一个家庭。

荷生问：“令堂住在什么地方？”

烈火露出一丝笑容，“问得好。”

爱一个人的时候，提起他，才会有笑意，烈火很明显地爱母亲，孝顺孩子坏不到哪里去。

“家母一半时间住伦敦，另一半时间住新加坡，几时我们去看她。”

“你外公也是财阀？”

烈火答：“不，家母娘家清贫，慢慢我告诉你。”

荷生点点头。

这时候烈云走过来，笑嘻嘻地看着他们。

荷生把她拉过来，搂住她。

三个人都没有发觉，烈风躲在一株树影下，正静静地留意他们的一举一动。

跟着的一段日子，要求再苛刻，荷生也得承认她对感情生活极为满足，趁着母亲到加拿大度假，不必准时回家报到，她与烈火争取每一个见面的机会。

荷生诧异时间飞逝，似有人故意拨快时钟来作弄他俩，由拂晓到黄昏，又自黑暗到黎明，一瞬即逝，不可思议。

其间也发生过一两宗不愉快的事件，荷生不能不耿耿于怀，却不得不加以容忍。

一次她与烈云吃茶，碰见言伯母。

彼时言诺与烈火在日本开会，自分手后，荷生始终没再有机会见到言诺，但对言伯母的尊敬仍在，故此特地过去打个招呼。

荷生的生活经验不足，满以为是一番好意，谁知言伯母不领情当众奚落她，当时她上上下下打量她，似要重新估计荷生，然后冷冷地说："对，夏小姐，同伙计的儿子走不如同老板的儿子走。"

荷生年轻面皮薄，没想到一向和颜悦色的言伯母翻起脸来会如此惊人可怕，急急回到自己坐位，忍不住落下泪来。

也是应该的，她怎么可以妄想言家像以前那样对她？

不能怪言伯母小气。

说也奇怪，挨骂之后，荷生觉得恍惚补偿了什么。

但是过了三天，荷生又发觉她连这样卑微的安乐都得不到了。

烈火告诉荷生，言诺为这件事连声道歉，荷生可以想像他一额汗满脸通红的样子。

荷生问："言诺为什么不亲自跟我说？"

烈火没有回答。

荷生有点气，言诺没有怪责烈火，但是他不原谅她，男人与男人永远有默契，因此女人永远成为罪魁祸首。

“他请你体恤一个母亲的心情。”

荷生无话可说。

烈火说：“相信我母亲会做同一样的事。”

这样的小事，荷生愿意忘记。

比这大一点的事却令荷生紧张：烈火替她安排好，过两天去见他的母亲。

养这样漂亮儿女的母亲自然是美人。

荷生相信她不会失望。

她满以为可以同时见到烈先生和夫人，但是接待她的只是烈太太。

荷生一早听说，新加坡有一群华侨受英国文化影响甚深，比香港人有过之而无不及，看到烈太太的排场，荷生完全相信这个说法。

银茶壶盛着格雷伯爵茶，青瓜三文治，小小软面饼夹奶油及玫瑰果酱。

烈太太穿着一套低调的丝质见客服，简单的首饰，看上去令人觉得舒服。

烈太太的脸同烈云一个印子，但气质过之，荷生相信她另有更大的优点令烈先生欣赏。

烈火过去轻吻母亲的脸颊。

然后他退出去，让两位女士清谈几句。

烈太太轻笑，“我们期待看烈火的女朋友已有多年。”

荷生笑一笑。

“欢迎到烈家来。”

荷生处之泰然，大大方方地说：“我来过好几次了。”

“那么，你大概已经发觉，这幢屋子，叫琪园。”

荷生一怔。

一说就说进这种题目上，看来，已经把荷生当自己人了。

“琪，是周琪，烈先生的前妻。”

荷生不敢露出什么蛛丝马迹。

烈太太声音有点无奈，“我本人姓陈，叫陈珊。夏小姐，试想想，陈珊住在周琪的屋子里，那种感觉应是如何？”

她仰起面孔，尖尖的下巴单薄俏丽一如烈云，荷生觉得她太过介意这种细节，难怪神情忧郁。

荷生大胆地说：“或许可以把大厦的名字改掉。”

“烈先生不会这么做。”

“为什么？”

“大厦由他恩师所盖，他要纪念他。”

从这句话，荷生又知道得多十点，像是缓缓又推开一扇门，看到出乎意料的景色。

“夏小姐，我想说的是，烈家是一个奇怪的家庭，你要给烈火多些时间，多点耐心，同时，你本人需要有最大的弹性，去适应你从前没有接触过的问题。”

荷生不敢认同她语气里的悲观，为了礼貌，不予置评。

烈太太随即高兴起来，“烈火常跟我说起你。”

荷生这次知道如何应对，“烈太太同烈火看上去似姐弟，感情一定很好，什么话都可以说。”

试过多次了，没有不灵光的，无论诸位女士的外形出身学识如何，说她比她的真实年龄年轻，她一定开心，连这么聪明漂亮的烈太太也不例外。

荷生知道她做对了。

她比先前镇静，喝一口茶。

“烈云也很喜欢你，”烈太太停一停，“我这两个孩子都没有知己朋友，”那种无奈与忧郁又来了，“同我一样。”

荷生笑道：“烈云是比较内向。烈火有言诺。”

烈太太用手撑着头，过一会儿说：“对不起，夏小姐，我竟有点累，我们下次再谈吧。”

荷生连忙站起来，“当然，我先告辞。”

走到会客室门口，荷生还听见烈太太细细长长的一声叹息。

荷生走到走廊，口腔内还有茶的苦涩味，她正在想，下次喝茶，或许应该恢复加糖的习惯。经过书房，却发觉烈云伏在烈火肩上，正在饮泣。

看到荷生，兄妹俩立刻松开，烈火说："母亲责备她呢。"

不，这里边另外有个故事。

烈太太已经给她警告，有些问题，不是外人可以接受的。

荷生静静坐下来。

烈火问："母亲跟你说什么？"

荷生微笑，"只怕烈太太对我印象普通。"

烈火答："过些时候，她也不再是烈太太了。"

荷生讶异。

"她要与父亲分手，带走烈云。"

噢，所以哭泣。

烈火感慨地说："这间屋子，除去父亲，没有人会住得安乐。"

"过来，"荷生叫烈云，"坐到这边来。"

烈云心智非常弱小，遇事完全没有对策。

荷生搂着她问："你愿意跟父亲还是母亲？"

谁知烈云厌恶地说："我要自由，为什么我不可以选择，为何我不能同我喜欢的人在一起？"

荷生正在错愕，烈火忽然抓起烈云，劈头盖脑地给她一个耳光，用力甚大，把烈云的脸打得扭曲地偏过去。

荷生惊得呆了，下意识伸手去阻止烈火再出手，但是烈火已经紧紧抱住了烈云。

他悲痛地说："小云，小云，别让我伤心。"

荷生发觉外人简直没有插足余地，她悄悄站起来离去。

走到走廊，迎面而来的，却是言诺。

蓦然在陌路与他相逢，荷生睁大双眼，不知该做出什么反应。

言诺百感交集，贪婪地打量荷生，她瘦了，心事重重楚楚可人，他闭上眼睛，似怕她炙痛他双目。

荷生要再次见到言诺才知道，她同他在一起的时候，并非不快乐。

两人僵持一会儿，荷生听得言诺轻轻地问："好吗？"

荷生想说好，只觉不妥，想说不好，实在又太草率。

她呆呆站着。

就在这个时候，烈火叫着找出来，"荷生，荷生！"他终于发觉她已离开书房。

看见她与言诺对望，烈火笑问："荷生，你怎么不声不响走开？"

荷生抬起头，"呵，我不过想到花园走走。"

言诺也连忙笑道："再不浇水，你那些灵芝草野山参快要枯萎了。"

三个年轻人要这下子才明白，什么叫做强颜欢笑。

言诺说："烈先生在楼上偏厅等我。"

烈火说:“一起吃饭吧?”

“不，你们去，我恐怕要讲一些时候。”

烈火笑，“父亲从来不同我商量这样的大事。”

言诺匆匆走上楼梯。

荷生惆怅地看着他的背影。

在烈火眼中，也就是旧情绵绵，女孩子就是这点古怪，她们记忆力太过惊人，好似什么都不能忘记，一生中所有的琐事永存脑海，一有风吹草动，便拿出来回忆一番。

他没有耐心地等荷生回过神来，他问:“刚才你想走到哪里去?”

荷生答:“我看不惯兄弟姐妹动辄用武力攻击对方身体。”

烈火说:“在这间屋子里，我们只能做到这样。”

“胡说。”

“将来你会明白。”

“不，我永远不会了解。”

烈火指着玄关天花板的水晶灯说:“父亲在十五年前带着我同烈云来接收这幢房子的时候，就跟我们两兄妹说，世上没有什么是一个人应得的，一定要靠双手去争取。夏荷生，要是你看不过眼，现在还来得及。”

荷生握着拳头瞪着他，烈火一转身消失在走廊尽头。

他们俩的价值观显然有着太大的分歧。

荷生像是看到童年的烈火已经被训练成一只鹰，由父亲带进来做琪园的新主人。

这间宅子的旧主把毕生武艺与独生女儿都交托给一个野心勃勃、没有良知的年轻人，这人负了师傅一片苦心，这人夺产弃妻，这人是烈火的父亲。

荷生震惊之极，把所有的资料串在一起，她竟得到一个这样的故事。

可怕的漩涡。

所有不懂得维持安全距离的人都有机会被卷到激流中心溺毙。

荷生抬起头，那盏庞大华丽古典式样的水晶灯似要压向她头顶，她不由得后退两步，原意想靠在墙上，谁知却碰到一个人。

荷生连忙道歉。

人家已经伸出手来扶住她。

荷生穿着短袖衣裳，与那人在毫无心理准备之下肌肤相触，不由得轻微地痉挛一下，她下意识保护自己，抱着双臂，看向那人。

那人认得她，“是夏荷生小姐吧？”

他是个中年人，穿深色西服，有一股雍容之态，眉宇间像煞一个人。

荷生的心一动。

那人笑着自我介绍，“我是烈战胜，烈火的父亲。”

荷生虽然已经猜到三分，但听到他道出姓名，仍不禁有震荡感，是他，是这个人，忘恩负义，有才无德的衣冠禽兽。

荷生原先以为这么残酷的人应有丑陋的外形，但是他却温文尔雅，和蔼可亲，荷生错愕了。

烈战胜任由这个美丽的女孩子瞪着他。

幸亏言诺下楼来，“荷生，你见过烈先生了?”

荷生连忙说：“是。”内心犹自忐忑。

言诺又说：“烈先生，这是我们的朋友夏荷生。”

我们的朋友，这五个字简直可圈可点。

烈战胜早就见过夏荷生，那日在冷气车厢里，他亲眼看到烈火与言诺两人争相讨好这个女孩子。

那天，天气炎热，三个年轻人似在汗里捞起来，他们却丝毫不觉不适，谈笑自若，顾盼自如。

当时烈战胜感喟地想，年轻真好，即使一无所有，拥着青春，已经足够。

他注意到烈火的神情，知道他对这女孩子已经倾心，那时，也许烈火还不能肯定自己的心意，但是在有经验的旁观者眼中，三角局面已经十分明显。

能叫不羁的烈火为她同惟一好友言诺起冲突，这女孩的魅力也就很惊人，如今言诺做了败家，却不怀恨，可见她有过人之处。

烈战胜的慑人目光迫使荷生转过头去。在烈

宅，她一向有被偷窥的感觉。

难道一直是他？

荷生定定神，“我要走了。”

言诺意外，“你不是约好烈火了吗？”

荷生涩笑，“烈火今日情绪不好。”

烈战胜说：“这是他最大的弱点，真要他的朋友多多包涵。”

言诺有点为难，他送不送夏荷生呢？

避得过一时避不过一世，算了，问心无愧，何必避这个嫌疑，于是他说：“我送你下山。”

荷生吁一口气，礼貌地与烈战胜道别。

在车里，言诺温和地问：“吵架了？”

荷生当然听得明白，怅惘地答：“习以为常。”

言诺有点难过，他从来不与荷生吵嘴，他一向忍她。

半晌荷生问：“好吗？”

言诺点头，“非常忙，烈先生有意把我训练为父亲的接班人，家父想在短期内退休。”

荷生忍不住问：“言诺，烈战胜是否不道德地夺取了周家的财产？”

言诺看她一眼，避重就轻，“每一个成名的人，都会受若干传言困扰。”

“烈战胜可怕吗？”

言诺答得很含蓄，“就算是，我们也尚无资格看到他阴暗的那一面。”

“我觉得大家狼狈为奸，对付烈风，甚至是不遗余力。”

言诺说：“烈风是个悲剧人物。”

讲得再正确没有了。

“他父母之间官司诉讼十余年，烈风自幼至今便只知道父母是仇敌，先是离婚官司，然后是遗产纠葛案，他母亲输得一败涂地，连带把他也当筹码输了出去。这些年来，谁也没给他好脸色看。”言诺叹一口气。

荷生忽然说：“除去烈云。”

言诺吓一跳，连忙顾左右，“看我，说起是非来竟津津有味。”

荷生有感而发，“言诺，你最幸福。”

言诺一怔，这话竟出自荷生的嘴巴，太可怕了，荷生不知道她此刻的强烈优越感有多像烈家的人。

当下言诺尽是微笑，他问她：“真的吗？失去夏荷生，我还应当快乐？”

荷生闭上眼睛叹口气，“对不起。”

言诺停下车来，“替我问候伯母。”

最大方他也只能做到这样。

回到家，荷生倒在沙发上。

夏太太问：“那是小言吗？”

荷生点点头，走到厨房，拉开冰箱，捧着冰淇淋盒子，打开，就用调羹勺来吃。

夏太太有点惊喜，“你们言归于好了？”

荷生抬起头来，“不，但我们仍是朋友。”

夏太太感慨，“做人越来越难，多尴尬，还得硬着头皮。”

对，以前分手后可以名正言顺地交恶弹劾，尤其是女方，再失态也能博得同情，世界不一样了，现在要大方可爱地处理这种事……也难怪生癌的人越来越多。

荷生吃光一整盒冰淇淋，消化系统凉飕飕，她叹一口气。

“你与小言在一起的时候比现在开心。”

不，这并不正确，母亲没有看到她狂喜的时刻，她舍平淡而取激情，当然要付出代价。

“这次见面，你姨丈说，小饭店要人帮忙，叫我过去站柜台。”

“那多好，你的意思怎么样？”

“我？”夏太太看着女儿，“此刻走，总像放不下心。”

荷生何尝不明白，母亲关心的是她终身大事，但嘴里只说：“改变一下环境，半年后不喜欢再回来。”

“所有移民都高估自身的适应能力，冲动地走，悔恨地返，结果两边不到岸。”

荷生多心了，这是说她吗？

与母亲都生分了，可见这些日子她是多么的孤独。

没有人支持她。

深夜，电话铃响。

夏太太听见，不加理睬，荷生放下小说，去接电话。

烈火在那头说："荷生，让我们结婚吧。"

这也许是解决问题最好的办法：成为烈家一分子，同流合污，共享富贵。届时，再看不过眼的事情也都顺眼了。

"你还生气？"

"咎由自取，气从何来？"

"荷生，你总令我惭愧。"

荷生"扑哧"一声笑出来。

"出来，我载你兜风。"

荷生不能抗拒这样的邀请，她换上窄身衣裤，轻轻蹑足出门。

夏太太不是没听见，但无意阻止女儿，少年不胡作妄为大胆放肆，试问老年时哪来的题材话当年？其实她一直姑息荷生，只是荷生不知道。

夏太太看一看闹钟，清晨三时整。

荷生飞快地奔下楼梯，头脑是清醒的，一边同自己说，夏荷生，你像烈火名下小叭儿狗，逃不过他的魅力五指山，为什么这样听话，连叫他等一等都不舍得？

心底虽然不值，脚步却加速自四楼一直转下去转下去。

新月如钩似的挂在天边，烈火的大型机器脚踏车咆吼着驶过来，拐一个弯，停在荷生面前。

烈火戴着头盔，浑身漆黑，带种妖异的感觉，他把后座头盔交给荷生。

荷生熟练地坐上机车，跟随烈火飞驶而去。

她不知道他要把她带到什么地方去，她也无所谓不在乎，只要与他在一起，到哪里都是一样。

速度本身已是最大的快感，荷生闭上眼睛，愿意永远这样同烈火逍遥自在放荡不羁地奔向永恒。

车子终于停下来。

烈火摘下头盔，笑嘻嘻地看着她。

“跟我一起之后，怕没有人敢要你。”

荷生笑，“怎么见得我会要别人？”

机车停在一间小小别墅门口，荷生一看便喜欢，庆幸它不是西班牙设计，而是朴素的英式平房。

烈火掏出钥匙，想开启大门。

一推，不开。

烈火一边说：“这个地方最清静，本来是家母闲时来写生用的。”

“伯母是画家吗？”

烈火用力推一推门，“嚷，难道带错钥匙了？”

荷生很客观地说：“里边下了锁。”

烈火笑，“里边没有人。”

“清洁工人呢?”

“不留宿。”

“那么一定是自己人。”

谁知烈火即时变色，“荷生，站在大门口不要动，我到后门去看看。”

荷生拉住他，“我们走吧，假如屋里有人，碰上面也不好。”

他们何尝不是来幽会的。

“你别管。”

烈火一径奔到后边去。

他一走，前门马上打开门，一个影子冲出来推开荷生，力量甚大，荷生十分狼狈，踉跄两步，差点摔跤，那人逃出大门，狂奔而去。

那人穿着黑衣黑裤，头上压着一顶帽子，又用一方面巾捂着嘴。

但是没用。

荷生已经知道他是谁了。

她扶着墙壁站定，满腹疑虑：他来干什么?

这间小别墅明明是烈火母亲的私人产业，这个人怎么会有钥匙?

荷生听见远处传来汽车引擎声，那人开车走了。

荷生惊疑地推开大门，看见楼梯底坐着一个人。

没有开灯，荷生一时看不清楚，那人缓缓地抬起头来，借门口的微弱光线，荷生发觉坐着的是烈云，她赤脚，身穿浴袍。

荷生这一惊非同小可，她吓得张大嘴巴，不知所措。

过半晌才能问："烈云，是你？"

烈云站起来，脸色雪白，"求你别说。"

这时烈火的声音已在她俩身后响起，"谁在屋里！"这是一声暴喝，分明震怒已极。

荷生只得握住烈云的手，转过身去说："是烈云。"

"谁从这里开车离去？"

烈火一手拨开荷生，对着烈云。

荷生一连被推两次，不禁心中有气，也大声说："是一只红颜绿头发的鬼！"

烈火一呆。

荷生再一次挡在烈云身前。

烈火责问："你亲眼看清楚是个外国人？"

荷生冷笑一声。

"叫什么名字？"

烈云这时镇定下来，"阿尊。"

烈火自喉咙底哼出来："我不信！"

"阿积。"烈云已经奔上楼去。

烈火要追，荷生挡住他，"她不是小孩子了。"

烈火看到荷生的眼睛里去，"你肯定看到的是外国人？"

荷生并无惧色，也瞪着他，“你对烈云客气点好不好?”

烈火缓缓坐下来，不出声。

“我们走吧。”

烈火不动。

“下次要用地方事先约好时间，免得无趣。”

荷生握住他的手，用力把他拉起来。

他们俩败兴而返。

时近拂晓，雾大露重，荷生心头载着一桩大秘密，忐忑不安，神情呆滞。

烈火放她下车，荷生伸手过去，轻轻触摸他面庞。

她轻轻说，“我们改天再去。”

烈火吻她的手背，不发一言上车离开。

悄悄返回屋内，关上门。

荷生知道母亲知道她的行踪，夏太太也知道荷生知道这一点，母女俩都不拆穿，都让对方以为已经成功地瞒天过海了。

谁会真的爱上做戏，不过是为了好下台。

荷生皱着眉头躺在床上。

过很久才睡着。

一下子天就亮了，偏偏她有一节课在上午九点，荷生撑到课堂，眼底发黑。

一边抄笔记一边用手托着头，每写每错，结果连自己也不耐烦起来，在笔记本上打一个大交叉，一手把本子扫到地上去。

她深深叹一口气。

下了课，走出课堂，已经有好事的同学笑着说：“夏荷生，有人在钟楼下面等你好久好久好久好久了。”

荷生讶异，自己也急于看个究竟。

身后还传来闲言闲语：“怪不得功课退步得那么厉害。”

“太值得了，荣誉生不值一哂。”

“真的？我们快去看。”

荷生奔下楼梯，见是烈火的车子，知道事情与人们想像的有点出入。

他从不骚扰她的功课。

荷生走向前问：“烈火，什么事？”

烈火转过头来。

荷生意外地笑，“你把胡髭刮掉了。”

他却无暇同她说这些，“烈云发高烧，今晨进的医院，她口口声声说要见你。”

荷生不假思索，“好，我们马上去。”

“谢谢你。”

荷生拍拍他的肩膀。

医院就在大学堂隔壁。

烈云在病房内昏睡。

看上去可怕极了，细软的头发搭在额上，脸白如纸，嘴唇颤动着。

荷生过去握住她的手，小云虽无知觉，却本能地握紧手指，渴望接触。

荷生不忍轻声地问烈火：“令堂呢？”

“她走了。”

“她不是要同小云一起走？”荷生意外。

“烈云不愿意走。”

哦，烈战胜又战胜一次。

烈云手腕上缠满针管，额角不住沁出冷汗。

多么奇怪的一个女孩子，忽如仙女，忽似修罗。

无论怎么样，她都使荷生心痛。

烈云呻吟一声，睁开眼睛。

荷生连忙转过头去，“烈火，帮我买一杯咖啡好不好？”

烈火出去。

荷生把耳朵附在小云嘴边，“现在房里没有人，你有话，尽管对我说好了。”

烈云张嘴无声，只是流泪。

荷生心酸，“你放心，我不会说出去，这同我有什么关系，我要来坏你的名誉？我发誓，要是我泄漏一言半语，叫我嘴里生癌。”

烈云眼泪汩汩流下。

荷生替她擦干泪水。

“把身体养好，还有大把日子要过，烈火同我都放心。”

小云点点头，她已经力竭，转过头去。

“不要理我们，你睡吧。”

她闭上眼睛。

烈火推门进来，“这里没有卖咖啡的，我们呆会儿出去喝。”

荷生站起来，“好。”

看护说：“让她休息吧，晚上再来。”

烈火与荷生并肩走到楼下。

“小云一遇惊吓，就会发高烧，自幼如此。”

荷生无语。

“告诉我，从别墅走脱的到底是谁？”

“我已经告诉你了。”

“你撒谎。”

“别太武断。”

烈火咬牙切齿地说：“你不说我也知道是谁。”

“那又何必来问我？”

烈火既怒又伤，“荷生，你到底站在哪一边？”

“对不起，烈火我没有愚忠。”

烈火也觉悲哀，“荷生，为什么我俩当中夹着这许多人与事？”

荷生答：“环境给我们什么，我们就得接受什么。”

烈火把脸埋在荷生双手里，“我或许不该把你自言诺怀中抢过来。”

“啊，有人后悔了。”荷生故意轻松。

“后悔？永不，我只是怕你吃苦。”

荷生微笑，“谁都知道我的物质生活比从前丰足，但是精神备受困惑。”

“不足以补偿你的损失。”烈火说。

荷生惘然，一时不知男友说得对不对。

回到家中，看见桌面上放着一张象牙白色帖子。

打开一看，荷生呆住，请夏荷生拨冗光临的人，竟是周琪女士。

荷生实在忍不住，找到言诺，开口便说：“烈风的母亲要见我。”

言诺沉默半晌，才说：“不要去。”

“为什么？”

“如果你征求我的意见，我劝你到此为止，一个人知道得太多无益。”

“言诺，你知道得比谁都多。”

“但我不是烈火的女友。”

荷生不出声，言诺当然有怨怼。

小言再次提出忠告，“同他们家人保持距离为上。”

“我用什么借口推托？”

小言叹一口气，“用推我的同一方法。”

荷生问：“我们不能做朋友吗？”

“我不会对陌生人讲这么多话的。”

“谢谢你，言诺。”

荷生没有接受小言的劝告。

他不再像以前那样，同她讨论、商量、提出建议，然后一起做出结论，用其中最好的办法。

他仍关心她，但是维持隔膜的距离。

车子来接她的时候，荷生准时去赴约。

大家即大家，周女士并没有要客人等。

她迎出来，烈风站在母亲背后，苍白瘦削，如一块褪色的布景板。

周女士让荷生坐。

荷生只觉此情此景何等熟悉，想转来，原来她接受陈珊女士招待的情况尚历历在目。

烈战胜的大夫人要比二夫人沉着老练。

荷生喝一口茶。

涩味中带点清香，两边府上仿佛用同一种茶叶，味道非常特别。

周女士坐在一张安乐椅上，烈风一直站在她身后。

她说："夏小姐，多谢你赏光。"

荷生欠一欠身子。

她又说："像你这般人才，同烈火这样的人在一起，实在可惜。"

荷生不由得扬起一条眉，他们竟斗得如此白热化，不替对方，亦不为自身留一点点余地。

周琪女士有一张尊贵的长脸，细狭眼睛，薄薄嘴唇，颇似中国历代帝后像中嫔妃的相貌。

烈云同她母亲的长相无异，较为俏丽。

“烈风说，你对他很客气，对他好即是对我好，所以请夏小姐来面谢。”

“呵，他对我也一样。”

“夏小姐，你是琪园的常客？”

“去过数次。”

“琪园，是一九四九年，家父为我盖的房子。”

荷生点点头。

“但是我却不能住在琪园内。”

荷生词穷，总不能安慰她说“一个人吃多少穿多少是注定的”吧。

言诺永远是对的，她的确不该赴会。

“家父与我都看错了烈战胜，我俩有眼无珠，好比盲人，应遭此报。”

荷生听周女士说得如此怨毒，不禁劝道：“依我看，这间屋子，比琪园更新式更舒适。”

她一怔，笑了，借口退下。

在这样的环境底下，再好的菜式也于事无补，荷生吃得很少，烈风拿着一杯白兰地，沉默地坐着陪客。

荷生怀疑烈家从无喜事。

烈火能够这样开朗实在不易，荷生心头一暖。

没想到烈风忽然幽默地说：“气氛不算热烈是不是？”

荷生笑。

烈风凝视她，“烈火这人，什么都没有，就是运气好。”

荷生问：“这是对我的褒奖吗？我打算照单全收。”

“你受之无愧。”

荷生轻轻地说：“或许你可以尝试解一解父母之间的死结。”

“名为死结，如何能解？”

说得极是，荷生觉得烈风的聪明比烈火有过之而无不及。

“或许你应该从头开始。”

烈风喝一口酒，“那个时候，我还是儿童。”

“对不起。”

“没关系，你算得是半个自己人，凡事何用瞒你。”

“那么，能不能把结怨的过程简单地说一说。”

烈风抬起头，像是在整理故事的段落，良久开不了口，可能事情实在有点儿复杂，他不知从何说起。同时，烈风亦颇为诧异，他一直以为言诺或烈火，甚至是两人一起，早就把故事说给夏荷生听过，且无可避免地丑化了他们母子这一方。

但是看荷生的神情，却明明未知首尾，烈风意外。

过一刻他才开始：“烈战胜同家母婚后一直在周氏机构身居要职，野心勃勃，对我外公阳奉阴违，对家母不忠不实，在外早有新欢。”

烈风直呼其父姓名，不予丝毫尊重。

“烈战胜终于等到机会，十三年前，我外公出事，涉嫌一宗行骗案，被控拥有空壳公司，无足够抵押向银行贷款，与案有关的串谋朱某是银行副主席，一直是周氏的好友，猜一猜，努力指证两人行骗的是谁？”

荷生不忍听下去。

“是烈战胜，”烈风说，“我的父亲。”

荷生闭上眼睛。

“老人在案子结束之前心脏病发逝世。再猜一猜，他把大部分财产送给了谁？”

荷生低下头。

“又是烈战胜。家母觉得老人立这样的遗嘱只有两个可能：一，他遭受恐吓；二，他神经错乱。于是聘律师起诉，但她没有赢得官司。”

荷生忽然觉得疲倦及口渴。

“接着，烈战胜与家母分居，随后单方申请离婚。他又如愿以偿。从此之后，他不正眼看我。我失去长子应有的名分地位，烈火取代了我的位置。假使你是我，你会怎么想？”

荷生叹口气，低声说：“我恨他。”

“对，我恨他。”

之后，烈风不再说话，他自斟自饮，荷生冷眼旁观，却不觉得他比稍早时更醉。

烈风的故事令荷生不胜负荷。

她站起来告辞。

烈风让司机送她回去。

在门口，荷生做最后努力，“烈风，忘却往事，从头开始。”

烈风站在晚风中，很温和地回答：“人一旦失去曾经拥有的矜贵身份，不容易放开怀抱，也不会甘心愿意那么做。”

荷生无言离去。

没想到会与烈风成为朋友，烈火要是知道，反应一定激烈。

荷生返到家中，见母亲外出，屋内静悄悄，并无倾诉对象，便卸妆洗脸，做了冷饮，喝个饱，正欲休息，忽然听到有人叫她。

“夏荷生，夏荷生。”

她抬头问：“什么事?”

两个黑衣妇人不知几时已经不请自来，一人一边，拉扯荷生，“快，快，周老爷快要归天，你还不随我们来?”

荷生才要辩说不认得周氏，已经被她们拘着越走越远。荷生嚷：“慢着，我要同母亲说一声。”

妇人们笑说：“夏太，早就知道了，你以为她是胡涂人?”

荷生只得跟着她们走，脚步如飞，如腾云驾雾。

一下子来到琪园，游上二楼，妇人对着一扇门说：“还不进去。”用力一推，便把荷生推进门去。

荷生只觉身体毫无困难地穿过大门，来到房内，还在讶异，只见房内黑压压的站满人，房中央一张大床，床上躺着一位老人，正在呻吟。

荷生下意识地知道，这人便是周老爷——周琪女士的父亲，烈战胜的岳父，亦即是烈风的外公。

荷生看到周琪跪在床头握紧父亲的手，像是在恳求宽恕，奇怪，她看上去好年轻，烈风呢，荷生的目光搜索烈风，呵，他照例站在母亲身后，怎么，还是个少年哪，荷生惊讶，灵光一闪，才明白她回到多年之前去了。

荷生想叫出来，但看见老人吃力地挥手，“去，走。”他要逐开周琪。

这是怎么一回事儿？

老人接着示意烈战胜过去。

荷生看到周琪恨恨地退开。

老人当着医生、看护、律师的面说：“我已立遗嘱……”说到这里，脸色已变。

荷生害怕，退后两步。

周琪站在角落，脸色阴沉，握紧拳头。

荷生像是明白了什么，她问周琪：“是你，是你辜负了周老爷？”

周琪却没有听见，拉开房门就走，荷生不由自主地跟出去，走廊又黑又长，走来走去看不见亮光，走来走去见不到尽头。

荷生惊恐已极，大声叫喊，一跃而起。

哪里是琪园，她躺在家中沙发上魇着了。

窗外淅淅地下着秋雨，十分富有情调，荷生见露台外晾着衣服淋湿未收，连忙去把衣架子抬进室内，一忙，把梦境忘掉一大半。

烈家的人可不让她喘息，电话急随而至。

烈火对荷生说："小云的情况已得到控制。"

这倒是一个好消息，荷生松口气。

烈火说："我俩许久没有私人时间了。"

"我要做功课。"

"本想教你做坏学生。"

"还用你教，我可以做你师傅。"

"万幸我比你早毕业。"

"对，别影响到言诺。"

烈火沉默一会儿，"关心他是应该的。"

"你多心？"

"你想。"

荷生那篇功课一直没有写好。

第二天她随烈火出海，快艇飞驰，阳光与浪花随风打在她脸上，黄昏回来，面孔晒得金光四射。

回到岸上，荷生都觉得身子左右隐隐摆动，如置身海浪，微微似有晕眩感觉，也是一种享受。

她累得走不动了，烈火把她背上四楼。

在门口碰见夏太太，烈火急急放下荷生，打个招呼，飞奔而去。

荷生知道她与烈火之间已经容不下其他事，包括母亲与那警戒的眼光。

荷生想搬出去住，又怕伤害母亲。奇怪，此时此刻，最重要的是与烈火在一起，荷生心中几乎没有别的念头。

荷生不相信她会变成这样，把所有的精力兴趣都集中在烈火身上。

多么危险。

最后交上去的那篇功课，是花三百块费用请同学捉刀做的。

书友中有一早具经济头脑的人才，很坦白地说："荷生，我写的全是行货。"

"不要紧，"荷生微笑，"趁真正救世的天才尚未出生之前，多赚一点稿费。"

他很愉快地说："真的，没有人好过我即可，我何用好过自己。"

荷生并不担心此君，荷生担心她自己，学期开始以来，尚未打开过书本，有不少课文需要死背，如何考试?

烈云出院那日，荷生没有随烈火去接，荷生怕她的出现会令烈云想起那宗不愉快的事，她洞悉太多秘密，她怕烈云不自在，烈云需要静养。

过两天荷生在琪园大门口碰到烈云。

"好吗?"荷生笑着招呼。

烈云转过头来，神情仍然有些恍惚，见是荷生，放下心来，便问："等二哥？"

荷生正坐在烈火的车子里。

"你呢？"

"我出来吸口新鲜空气。"

荷生下车与她并排散步。

是烈云先提起，"你见过周琪女士，也见过我母亲，觉得怎么样？"

荷生非常诧异，只有一个人能把这次约会的详情告诉她，荷生冲口而出："你还在见他？"

烈云牵牵嘴角，笑得苦苦的，"我只关心他一个人。"荷生失措，"烈云，这是不对的。"

烈云看着荷生，"什么是对，什么是错？"

"社会自有一套律例，虽未臻完善，我们亦应尽量遵守。"

烈云笑了，握住荷生的手，"你真的关心我？"

荷生点点头。

"那么我不妨告诉你，我知道你在想什么，但事实跟你看到的，颇有出入。"

"烈云，我想你还是同那个人疏远的好。"荷生急了。

烈云想要解释，踌躇了一下。

烈火已经出来，叫荷生上车。

荷生对烈云说："考虑我的劝告。"

那边烈火兴高采烈，"父亲早该下这个决定

了。”

荷生看他一眼，是什么决定令他如此开心?

烈火神采飞扬，“父亲到今天才肯把烈风逐出局。”

荷生的心一沉。

“从此之后，不让他踏进公司半步。”

荷生吃一惊，烈火恨他的兄长，远比恨一个陌生人多。

烈火转过头来对荷生说：“我希望父亲登报正式同他脱离关系。”

荷生说：“烈火，你已是你父眼中的苹果，早就是他的储君，何用逼人太甚。”

烈火看着女友，“今日心情太好，不同你争论，”他笑，“我们到什么地方去庆祝?”

他开动车子，荷生在倒后镜中看到烈云小小苍白的身形越来越远，越来越远。

荷生肯定她已听见刚才烈火的那番话。

烈火继续说：“父亲想同你吃饭，我替你约了星期三。”

荷生这才回过神来，“呵，那我要去置件正经衣服。”

语气与脸容不见欠喜气。

烈火得到的，正是烈风失去的。

荷生几乎想跑到烈战胜面前去说：“你的偏心造成他们兄弟阋墙。”

后患无穷。

身为父亲为什么要那样做?

“你看你，下次我再也不会把公事告诉你。”

公事，铲除兄弟叫公事?

当夜很晚很晚，烈云由言诺陪着上来找荷生。

夏太太去开门，先看到小言，心头一热，随即发现他身后的少女，以为那是他的新伴侣，热情又冷却。

荷生披着浴衣出来见客。

小言无奈地说：“小云逼着我带她来找你。”

荷生问：“什么事?”

小言识相地说：“你们到露台去商量吧。”

烈云说：“言哥哥我不介意你听。”

言诺苦笑。

烈云开口，“我不能坐视父亲同二哥联合起来对付烈风。”

荷生立刻说：“烈云，这种事你不宜夹在其中。”

“你还看不出?烈风是无辜的。”

“我也看出，你越帮他，烈火越恨他。”

言诺这个时候说：“荷生讲得好。”

“这么说来，他只有我了。”烈云相当镇定。

“烈云，我劝你丢开这件事，外边世界海阔天空，不一定要在琪园争一席之地。”

烈云看着荷生，“说时容易，你是外人，况且你很可能做琪园将来的女主人，你当然这样说。”

荷生无言。

言诺问：“你想荷生怎么帮你？”

“请她代为说服烈火放弃驱逐烈风。”

荷生叹口气，“你太高估我，在公事上，我一点力量都没有。”

烈云不置信地说：“二哥哥那么喜欢你。”

“你让他学猫叫学狗吠是一回事，小云，你认识你二哥，这种决策没有人可以影响他。”

烈云缓缓低下头来。

言诺轻轻地说：“你总算了解烈火了。”

小云站起来，“那么只好由我自己想办法。”

“烈云，我已经劝过他。”

烈云低声说：“烈风千方百计想继承他外公——”

荷生忍不住，“我有种感觉，小云，你一直越帮越忙。烈火不愿意你与他们接近，你为什么不明白？”

言诺要阻止荷生，已经太迟。

烈云脸色大变。

荷生叹一口气。

言诺说：“小云，我先送你回去。”

烈云看着荷生：“我以为你是我的朋友。”

“我的确是。”

烈云摇摇头，随言诺离去。

荷生几乎想捶胸尖叫出来。

烈家没有一个人肯往后退一步半步，统统坚持站在针尖上僵持，且把她作为核心。

荷生用手捧住头。

夏太太过去用手按住女儿的肩膀。

荷生问："母亲，我应该怎么做?"

"你舍得离开这个叫做烈火的人吗?"

"不可能。"

"那么别问。"夏太太说，"去休息吧，时间不早。还有，我已经申请移民，短期可望批准，去加拿大料理餐馆。"

"是几时的事，"荷生站起来，"为什么不告诉我?"

夏太太微笑，"你哪里还有空儿理这些。"

荷生已与外边世界脱节，如陷迷雾阵中，挽住烈火的手，便心满意足，看到他人安排生活，只觉营营役役，琐碎无比。她没想到，此刻的夏荷生受人操纵，已无自主，被牵着向迷宫中央走去。

传说迷宫中央都住着一个魔王。

荷生怀疑烈战胜会随时拉下面具，露出原形。

魔王有角、长尾、皮肤起鳞片，外形奇丑。

烈战胜却不是那回事，从远处看他，年轻一如烈火的大哥，表面功夫，又胜过烈火许多。

荷生整晚都没有看见烈云。

她关心地问起小云，烈火简单地答："今天没有见她。”语气中有跋扈专制的意味，荷生非常不喜欢。

荷生活泼起来可以相当投入，但这个晚上，她是个槛外人。

整个晚上，她只肯说“是”、“不是”、“过得去”、“不错”，烈火笑她如接受律师盘问。

饭后烈战胜说：“叫小云下来喝杯咖啡。”

烈火离开图书室，烈战胜便对荷生说：“夏小姐好像对我有点误会。”

荷生诧异，“你在乎别人怎么看你吗？”

烈战胜笑笑，“很多时候不。”

对了，这才像烈家主人，管他人满不满意，他是法律，他至高无上。

“我猜想有人对你说过我的故事。”

荷生坦白地点头说：“是的。”

“夏小姐，你那么聪明的人，应该明白，你听的版本，都只是对说故事人有益的版本。”

荷生笑笑，“你又不肯说。”

“我很少解释。”

荷生想，说不解释，已是解释。

“夏小姐，我在乎你的看法。”

荷生抬起头来，“为什么？”

“我有种感觉，你会留在我们家中颇长一段日子。”烈战胜目光炯炯。

荷生牵一牵嘴角，会吗？从现在到火焰熄灭，还有颇长的一段日子？连她自己都没有把握。

这时烈火下来说：“小云不在房内，她出去了。”

荷生帮着烈云，笑问：“你规定她每次外出都要向你报告？”

烈火看女友一眼。

烈战胜问儿子，“你有没有对夏小姐说过我们家的故事？”

烈火喝一口咖啡，“我们家有故事吗？”

荷生见他否认得一干二净，手法比他父亲还要老练，不禁骇笑。

看样子今天晚上的烈战胜的确有话要说。

刚要聚精会神听故事，荷生听得门外一阵骚动。

有人在走廊处争吵，烈火出去看个究竟，过一刻他进来说：“烈风要求见你。”很明显，烈风此刻被拦在门外。

烈战胜神色平静，“让他进来。”

烈火对荷生说：“我想你避一避。”

他父亲却道：“不用，荷生可以坐在这里。”

烈火扬声吩咐：“放他进来。”

荷生如坐针毡，唇亡齿寒，将来烈火失势，这些人还不知道要怎么对她。

烈风满面怒容冲进图书室来，他在走廊经过一番挣扎，衣领被扯在一边，气咻咻半晌出不得声。

烈火静静地坐在父亲身旁。

只听得烈战胜说："关上门，坐下。"

烈风尽量按捺怒火，照他父亲指示而做。

烈战胜又说："把你的来意扼要地说出来。"

烈风声音颤抖，"让我留在公司里。"

烈战胜一口拒绝，"我要服众，没有商量。"

"那是我外公周氏的事业，你不能胡乱找借口驱逐我。"

"烈风，你外公另有产业留予你。"

"他也答允让我在机构里占一席位。"

烈风紧握拳头，瞪着他父亲。

烈火缓缓站起来，留意着烈风的举动。

"这个决定对你的前途没有丝毫影响，烈风，我劝你去外国度假静思，别让你母亲左右你的行为。"

谈判完全失败。

烈风忽然狂吼一声，向他父亲扑过去，荷生本能闪避，烈火伸出手臂拦腰抱住烈风，荷生连忙开门招呼下人。

把两人拉开的时候，双方嘴角都挨了一拳，嘴唇破裂，淌下血来。

一个管家一个司机把烈风箍得紧紧的。

荷生过去说："烈风，我送你回家。"

烈火用手抹着嘴角，听见这话，吼道："荷生，不准你动。"

有人在门外说："那么，由我送他。"

众人转头一看，是烈云自外返回。

烈火冷笑，"小云，你疯了?"

烈云丝毫不惧，"是吗，就算我是疯子好了，幸亏我不是你的女友。"

烈战胜叹口气，"烈风，你走吧，别再惹事。"

烈风大叫："把我应得的还给我!"

烈战胜走近他，看到他双眼里去，"没有什么是你应得的，在这个家，你要什么，要努力赚取。"

烈战胜将手中酒杯用力摔向墙角，大踏步走了。

荷生同烈风说："我们走吧。"

"夏荷生，你胆敢同这个人再说一句话，我就不认识你。"

荷生也是个极端不怕硬的人，她对烈火说："也许从头到尾我都没有认识过你。"

荷生拉着烈云送烈风出门。

到了门口，烈风悲哀地说："你们俩回去吧。"

荷生强笑道："我是外人，我不要紧，最多以后不来琪园。"

烈云靠着烈风的肩膀饮泣。

荷生觉得冷，拉一拉衣襟。

"烈云，你回屋里去。"

小云说："我不要回去。"

烈风叹口气，"我自己会走，不用你们陪。"

烈云欲趋向前，荷生拉住她，看着烈风上车走了。

烈火缓缓地从树丛走出来。

荷生问："是你？你一直偷窥我们。"

烈火命令烈云，"小云，回屋里去。"

烈云却恳求荷生，"让我到你家去住一晚。"

"你是成年人，你有自由这样做，来。"

烈火喝道，"荷生你胆敢纵容烈云。"

"说呀，"荷生疲倦地转过头来，"说你要剥我们的皮，说呀。"

烈火呆住。

荷生指着他说："你不晓得这个时候的你有多讨厌。"

她把烈火撇在大门口，与烈云乘车离去。

烈云开车如腾云驾雾，只想快，在这方面，兄妹俩非常相似。

她把车子铲落山去，半途在避车处停住。

烈云幽幽地同荷生说："你得罪二哥，不怕失去他？"

荷生反问："这么容易失去一个人？"

"你知道他脾气。"

"那么，失去也只好失去了。"

烈云钦佩地说："荷生，你真强悍。"

"环境造人，少年丧父，从此把一切大事看淡。"荷生深深吁出一口气，"同你刚相反，看你多么骄矜，小小不如意，即时哭泣。”

烈云低下头来，“荷生，你对我真好。”

荷生微笑，“我也觉得是，这是我弱点，我疼女性，据说最没出息的女人才珍惜女同胞，应当互相倾轧，争取男性的欢心才是。”

烈云苦苦地笑。

“来，到舍下度一宵，试试做穷人的滋味。”

“荷生你这样说真叫我没有藏身之地。”

到达夏宅，荷生侍候烈云沐浴更衣，又把自己的床让出来。

她笑说：“放心，垫褥底下没有豆子。”

烈云叹口气，“只有你把我当小公主。”

“烈云，他们是他们，你是你，为什么不跟着母亲出外过新生活？”

烈云笑，“荷生，这下可逮住你了，责己也要严啊，你呢，你为什么不跟令堂到外国从头开始？忘记烈火这个讨厌的人诚属好事。”

荷生一怔，丢下烈火？她想都没想过，光是听烈云说起有这样的可能性，已经心跳。

“做不到吧，其实我们每一个人都为自身套上一副枷锁，紧紧囚在牢笼里，不能动弹。”

夜已深，人已静，两个女孩子压低了声音。

“烈云，我还是要劝你疏远一个人。”

“不，你错了。”烈云按住荷生的手。

荷生看着她，“那人明明是你同父异母的大

哥。”

“每个人都这么想，但是烈风不姓烈，他父亲不是我父亲。”烈云透露一个惊人的秘密。

荷生讶异地说：“我不相信，小云，你一厢情愿，他同烈火长得非常相似。”

“英俊的男孩子都是一个模子出来的大眼睛高鼻梁，我们没有血缘关系。”

“但他的母亲周琪明明是前任烈太太。”

“那是真的，不过烈风的父亲另有其人，这件事我一早就知道。”

“烈云，是谁把这宗秘密告诉你的?”荷生非常狐疑。

“烈风。”

荷生张大嘴巴，但心中一颗大石缓缓着地。

“烈云，即使没有血缘，感觉上也尴尬，为何一定要选烈风?”

“选?”烈云仰高头笑起来，“荷生，原来你比我还要天真，你以为我们真有权选择?”

这话说得很玄，哲理甚深，荷生细细咀嚼。

荷生紧张地问：“烈火可知道此事?”

烈云摇摇头，“不能告诉他，也不能告诉父亲，否则烈风更加没有地位。”

“你一定要同烈火说，”荷生握住烈云的双肩，“他憎恨烈风，一半是因为你的缘故。”

“不，荷生，你要答应我，今晚的话，不能传出去。"

"谢谢你，烈云，"荷生啼笑皆非，"这些秘密，一件件如大石似的压在我胃里，迟早穿洞。"

"我们睡吧。"

睡，还能睡？

荷生想哭。

但是黑夜自有它的一套，彷徨慌张的心受它安抚，渐渐平复下来，荷生的双眼犹如胶着似的，黏在一起，她终于在客床上睡着。

第二天醒来，发觉烈云已经离去。

大概是睡不惯，急着要回家补一觉。

荷生也不以为意。

昨夜听来的故事，只当梦魇中的情节，荷生把它搁在一旁，暂且不去理会。

夏太太同女儿说："烈小姐说，多谢你招呼她。"

"你看见她离去？"荷生问。

"嗳，她走的时候，约七点半左右。"

"妈妈，你应该叫我一声。"

"她说不用你送。"

稍后，言诺的电话来到。

"听说你硬是把烈云带走了。"

"我没有拐带她，言诺，你必定是听了烈火片面之词。"荷生没好气。

"你叫她来跟我说话。"

“她已经走了。”

“走？”言诺紧张起来，“去了哪里？”

“我不知道。”

“真不知还是假不知？”言诺的口吻已似质问。

“言诺，烈云是一个成年人，我不能拘禁她，”荷生光火，“她昨夜在我处留宿，今早起来离去，你何不拨到琪园去看看，也许她在家里睡觉。”

“荷生，你并不认识烈云，你不该担这种干系。”

“言诺，要是你昨晚在现场，你也会做同样的事。”

言诺叹一口气，“听说昨晚真的闹大了。”

“烈云不得不避开一阵。”

“你俩昨夜可睡得好？”

“不好。”

“你同烈火吵架了？”

“已经不是新闻了。”

“荷生，有时我替你担心。”

荷生的鼻子一酸，连忙忍住。

言诺也知道他不方便多说，“保重。”

荷生把头枕在双臂上良久。

烈火并未登门道歉，也许他认为他没有错，但是在这样的关系里，谁爱谁多一点，谁就会自动认错。

电话铃尖锐地响起来。

是他，是烈火认错来了。

“荷生，我是言诺，”他气急败坏，“烈云不在琪园。”

荷生安慰他：“也许在逛街，也许约了朋友，言诺，你不过是替烈家打工，不必做兼职保姆。”

言诺当然听出讽刺之意，一声不响挂断电话。

荷生觉得歉意，但无法控制情绪，早知这么吃苦，就不该逞英雄与烈火闹翻，坐立不安真正难受。

她撇开一切上学去，下课时四处张望，没人来接。

荷生坐在钟楼下石阶上好一会儿，太阳下山，天色渐暗，荷生只得打道回府。

她没想到言诺与烈火两个人在夏宅等她。

呵，道歉还要人陪着来？荷生讶异，接着又感慨，三个人很久没有约在一起见面了。

言诺先沉不住气，“荷生，烈云不见了。”

荷生一呆。

“早上七点半自你这里离开之后，没有人见过她。”

荷生说：“还不到十二小时呢。”

“烈云的体质比较差，她很少连接逗留在外边超过三四个钟头。”言诺掏出手帕来抹汗。

荷生微微牵动嘴角，这并非身体不好，而是生活习惯娇纵，反正有的是时间，上场完毕，自然要回家休息一下，转个班，换件衣裳，再接下一场。

烈火背着他们，一声不响。

小言又问荷生，“小云有没有跟你说，她要到哪里去?”

荷生摇摇头。

“她离开的时候，表情有无异样?”

“我并未目睹她离去。”

“她一定跟你说过什么。”

“言诺，你好像在审问我?”

言诺太忠于烈家，幼受庭训，他自然而然地跟着父亲的老路走，烈风说得也对，外人看来，烈火永远像主子，而言诺，不自觉地拜了下风。

其实烈家需要的是人才，不是奴才，言诺满腔热忱竟给旁人一个完全相反的感觉，十分不幸。

这样的形象一旦固定，他再也离不了烈氏机构，就像他父亲一样。

言诺接着说：“荷生，小云自你这里走脱，你要负一点责任。”

荷生见他一直钉着不放，便回他一句：“要追究责任，你还不是烈家的人。”

言诺十分震惊，他蓦然发觉荷生变了，她不再是那个听话的小师妹了。

他转过头去跟烈火说：“对不起，我不得要领。”

烈火说：“我知道她在什么地方，我会找她回来的。”

他拂袖而去。

言诺说：“我希望小云不是在烈风那里。”

荷生转过头去，“你一直劝我不要介入烈家的私事，现在轮到我提出同样的忠言，他们并不需要外人协助，这么些年都过了，不见得到今日才需要我同你来做诸葛亮。”

言诺低下头，过良久，才说：“我也是为朋友。”

但过分热心，便似只看门犬。

荷生说：“我们都猜对了，小云一定在烈风处。”

她站起来送客，言诺一时却没有离去的意思。

“听说你功课退步了。”

荷生莞尔，“是。”

“会毕业吗？”

“言之过早。”

荷生已经无话可说。

她肯定了一件事，时光倒流，她也会再一次离开言诺，现在她清楚知道他绝对不是她要的那个人。

荷生不再内疚。

“对了，”她说，“联络到小云，给我一个消息。”

言诺沉默一会儿，只得告辞。

夏太太自书房出来，“不再有复合的希望了?”

荷生诧异地问：“母亲你为何如此高估言诺?”

“我希望有人照顾你。”

“谁照顾谁还不知道呢。”荷生叹口气。

“你口气老练许多。”

“经一事，长一智，这几个月来我的确长高长大了。”

“那么，你认为同烈火做朋友是适当的选择?”

荷生笑笑。

夏太太吁出一口气，“也许这只是你们的游戏，倒叫我这个旁观者紧张得透不过气来。”

游戏?

荷生没想到母亲有这么幽默。

她把功课翻出来追补，在她这种年龄，读课文一目十行，永志不忘，书本页数刷刷翻过，念文科就有这点好处，荷生一下子温习到深夜，手边一卷巧克力饼干吃得只剩碎末。

电话铃响了，她母亲探头进来，喜悦地问：“改邪归正了?”

荷生索性让母亲高兴到底，“无论是谁，说我不在家。”

过一会儿夏太太进来说，“不管用，那位先生知道你没出去。”

是谁这么霸道?

人在不在家是另外一个问题，不愿意听电话被人逼着去听又是另外一回事。

“是烈火吗?”

“不，是他父亲。”

荷生感到十分意外，“噫，他找我有什么事?”

她合上书本，走到客厅，也不开灯，一拿起话筒，那边就说：“夏小姐? 我有急事要见你，请你立即下楼来。”

“烈先生你在何处?”

“府上楼下。”

“五分钟。”

荷生急忙将长大衣披在运动衫上，取过钥匙开门下楼。

烈战胜站在车子边等她，见到荷生，替她打开车门，荷生刚坐好，他便把一张字条交到荷生手中。

荷生知道非同小可，急忙打开看，字条只有三行字，用英文打出来，一眼就看通，荷生一下子像堕入冰窖里，双手颤抖。

烈战胜沉着地说：“你是最后见到烈云的人。”

荷生说不出话来。

“我已通知警方。”

“但是——”

“我生平不受恐吓。”烈战胜的声音沉着而镇定。

荷生再次摊开纸条阅读上面的句子：令嫒在我

们手中，切勿报警，赎人条款容后通知。

烈战胜低声说：“你无须内疚，小云最后出现在什么地方并不重要，但我希望你能提供线索。”

他把车子停在路边。

烈战胜出示一块布料，“小云是否穿着这件衣裳?”

料子在领口部分剪出，她早就注意到烈云只穿一个牌子的服装，昨天烈云换下衣服，由她替她接好，差点儿便要叹息有些人竟可花五位数字置一件常服穿着，没想到今天就发生这样的事。

“小云可有异样的表示?”

“小云平常的举止都一直是异常人。”

烈战胜不出声，过一会儿，他们背后驶来一辆车子，车头灯闪两下，停在附近，有人下车，走近来，俯下身子，出示警方证件，“这位是夏小姐吧，希望你能把当晚的细节说一说。”

荷生惊恐过度，呆着一张脸，出不得声。

“夏小姐，请你与我们合作。”

烈战胜忙道：“慢慢来，她同我说也是一样。”

便衣探员催说：“烈先生，我们要争取时间及线索。”

“我知道，被绑架的是小女。”

探员只得退下。

过一会儿，烈战胜低声问：“你可否把详情告诉我?”

荷生看着他，不知从何说起。

烈战胜取出一只银制扁酒壶，“喝一口白兰地。”

荷生佩服他的镇定，打开盖子，喝了一口酒。

“要是你愿意的话，我们可以找一个地方坐下来慢慢讲，不过，”他看看后面的车子，“他们会在附近。”

荷生终于开口，“烈云昨晚约在九时许来到我家……”

荷生有惊人的摄影记忆，心细如尘，烈战胜听了她的叙述，犹如亲自在场一般。

在这样的要紧关头，荷生仍替烈云隐瞒着若干秘密，由始至终，没有提到烈风这个名字。

稍后，荷生也明白到，她这样做，也并非纯粹为了烈云，在这种时刻，于烈战胜跟前，提到他所不喜悦的人，是十分不智的行为。原来，荷生感喟地发觉，她像所有人一样，不敢令烈战胜不高兴。

烈战胜听毕，对荷生说：“我现在送你回家休息，请勿跟任何人说起这件事。”

“烈火在哪里?”

“他在琪园等消息。”

“我能否到琪园陪他?”

“我认为暂时没有这个必要。”

“需要我的时候请立即通知我。”

“谢谢你。”

他的声音始终没有透露过一丝惊惶失措。

恐惧是会传染的，烈战胜一直保持着冷静。

他驾车把荷生送回家，看着她上楼，才缓缓离去。

荷生整夜对着功课发呆，天亮的时候，她把书本扫到地下，走到街上去散心。

马路上已有不少行人，匆忙间荷生只觉迎面而来的女孩子，个个都似烈云，荷生掩住脸，一个踉跄，险些摔跤，幸亏有好心人扶住她。

荷生睁开眼，见是一个穿白色校服的女学生，大眼尖脸，她紧握住人家的手，“烈云?”

那女孩错愕地甩开她的手离去。

荷生叹一口气，抢到一部街车，坐上去，关上门。

她对司机说出一个地址。

荷生想去那个地方看看。

计程车停在烈家那幢小别墅前。

荷生下了车，按过铃，没有人应，便兜到后园，轻易自厨房半开的气窗爬了进去。

屋内静寂一片。

三间睡房收拾得十分干净，荷生兜了一个圈子，回到厨房，做一杯茶，喝一口，坐下沉思。

忽然之间她听得有人在她身后问：“你也发觉有疑点?”

荷生整个人跳起来，茶杯“当啷”一声打得粉碎，裤脚上全溅湿，她转头一看，说话的人却是烈战胜。

“对不起。”他取过厨房的毛巾交给荷生。

刚才怎么没看见他？

烈战胜回答她的问题，“客厅左边还有一个书房。”

荷生借收拾遮掩尴尬。

“我打开前门并不见人，回到书房却又听到人声。”

荷生另外倒一杯茶，慢慢呷一口。

“你好像有话要说。”

“烈先生，烈云在家，生活得并不开心。”

烈战胜不出声。

“她有她的难处。”

烈战胜仍然不语。

荷生问：“昨夜可接到任何消息？”

“来，我给你看一样东西。”他站起来。

荷生跟他进书房，烈战胜指着书桌上一只小型电动打字机说：“你试打一下。”

荷生坐好，取过一张白纸，卷入打字筒，顺手打出“很久之前，有一位公主……”

荷生呆住了。

她不由自主，改变字句，打出“令媛在我们手中”。同样的字模，一式的字键，荷生记得字条中每一个字母的尖端都带一点点红色，同这部打字机二色带的效果一模一样。

荷生抽出纸，悬亮光处一照，水印透出厂商标志，同她看过的那张完全相同。

荷生张大嘴巴，那封勒索信，分明是在这间书房写成的。

烈战胜到这个时候，声音仍然刚强，只稍带无奈，“我是一个失败的父亲，没想到孩子的生活竟然这般不愉快。”

荷生静静地看着他。

“我已让烈火去销案。”

“昨夜一宿并无消息?”

烈战胜终于疲倦了，他轻轻摇头。

荷生已不觉得他有什么可怕，蹲下来，轻声说：“我相信烈云不会做这样的事来伤害你。”

“你好像了解她比我多。”

“世事往往如此，也许你了解我，比家母更多。”

烈战胜只得苦笑。

“给她一点时间，她冷静下来，自会出现。”

烈战胜脸色凝重，如说旁人把事情看得太简单。

荷生叹口气，她希望这只是一宗安排失当的私奔案。

门外有汽车引擎声。

荷生探头出去看，与言诺打个照面。

言诺如释重负，“原来你在这里，我们到处找你，差点以为失踪的是两个人。”

烈火跟在他身后，他无暇闲谈，匆匆走到父亲身边，低声讲了几句话。

烈战胜说：“那么，请荷生帮帮忙。”

荷生连忙问：“我能做什么？”

“烈云要跟你说话。”

荷生答：“没问题，什么地方，什么时候，我愿意去。”

时间安排在清晨两点，私人住宅区内一个公共电话亭。

电话亭边有一间二十四小时营业的便利店，夜阑人静，只有两个店员，没有顾客。

言诺把车子停在一边，问荷生要不要咖啡。

荷生看看钟，他们早到了大半个小时。

只剩便利店有灯光，似一格透明的盒子。

荷生接过纸杯，问言诺：“你有没有去过烈风那里？”

“烈风不在本市。”

“这资料可靠吗？”

“烈先生已派人二十四小时监视。”

荷生低下头，“言诺，我们能不能开诚布公地同那边谈一谈？”

言诺看着她，“由你做代表？”他揶揄她。

荷生不去理他，还有二十五分钟。

“对不起。”言诺又为刚才的话道歉，“我太鲁莽。”

“不要紧，这两天大家都太累太苦。”

言诺从倒后镜里看见，“烈氏父子到了。”他马上下车。

荷生坐在车里，直到喝完咖啡。

同车来的还有其他人，把一只小小录音机交到荷生手中，教荷生使用。

亭子里的公用电话在黑暗中响起，比预定时间早了五分钟。

荷生连忙拉开门，取过听筒。

公用电话亭里有一股不愉快的异味，荷生无暇理会那么多，开着录音机，贴住话筒，提高声线说：“我是夏荷生。”

那边没有回答。

“烈云，是烈云吗?”

“荷生。”的确是烈云的声音。

“烈云，你有什么要求，尽管说出来，一定答应你。”

烈云呜咽，“荷生，叫父亲救我。”

连荷生都忍不住说：“回家来，烈云，别再闹下去了。”

电话在这个时候“啪”地一声挂断。

“烈云，烈云?”

烈火拉开电话亭子玻璃门，“你听到她声音没有?”

荷生木着脸，把录音机还给他。

他递给父亲，荷生只听得烈战胜说了三个字，“付赎款。”

他们钻进车子，预备驶走。

荷生拉住烈火，“慢着，你没有把整个故事告诉我。”

烈火说：“现在不是讲故事的时候。”

荷生固执地说：“现在马上告诉我。”

这个时候，烈战胜忽然开口：“荷生，请到这边上车。”

荷生过去坐在烈氏父子当中。

车子驶出住宅区。

烈战胜沉着地说：“开头的时候，这件事只是一个游戏，烈云被邀请做女主角，她欣然接受，天真地一心一意要帮助一个人，心想事后最多被我放逐到外国几年，作为惩罚。”

烈火一直看着窗外。

烈战胜说下去，“她遭人愚弄了，主使人的目的是要利用她来伤害我，她中了圈套，”

荷生马上明白了。

烈火沙哑着喉咙说：“烈云如不无恙归来，我会杀他。”烈火紧握拳头。

荷生闭上酸涩的双眼。

她也被人利用了，从头开始，烈云便把她当一只棋子。

那么怯弱秀美的烈云。

荷生用手捂着脸。

这是一个连环套，夏荷生是最末的一个环节。

烈战胜看着她，“你的面色很差，荷生，回去休息吧。”

荷生颤抖的手拉住烈战胜的袖子，“我不该多管闲事。”

烈战胜转过头来，双目炯炯，“这件事与你无关。”

“烈云回来的时候，请通知我一声。”

荷生从来没有觉得如此寂寞过，放了学她就整日守在家中等消息。

一个星期不到，衣带渐宽，人憔悴，连她自己都讶异会瘦得这么快。

算一算，烈云失踪，已经有七天了。

第八日，下课，荷生在钟楼下看到比她更委靡的言诺。

荷生的心“咚”地一跳。

言诺说：“烈先生叫我来同你说一声：烈云回来了。”

“谢谢天。”荷生大力呼出一口气，拍着胸口，“不然我会难过一辈子的。”

言诺脸上没有喜色。

荷生觉得双腿乏力，坐倒在石阶上，“好家伙，以后我才不会再妄用我的同情心，言诺，你教训得好，不听老人言，吃苦在眼前。”

言诺静静地坐在她身边。

“小云是否自行返家?”

言诺摇摇头，“她被丢在一个废车场。”

荷生一怔。

“她坐在那里有好几个小时才被管理员发觉，通知警方，又隔了半日才领回家。”

荷生觉得不妥，“小云现在何处?”

“医院。”

“她受了伤?”

“没有表面伤痕。”

“到底是怎么一回事?”

言诺忽然握住荷生的手，“她竟不知道她是谁，荷生，她神志不清。”

荷生听到这个噩耗，张大嘴巴。

“荷生，医生说她可能不会痊愈，永远不再认得任何人。”

“不，”荷生号叫：“不!”

她撇下言诺，一直向前奔去，不知道要跑向什么地方，一直跑一直跑，奔到校园，筋疲力尽，倒在草地上，面孔埋在泥中。

言诺终于追上来，荷生颤巍巍站起来，伏在言诺肩膀上，放声痛哭。

接着好几天，荷生都没有烈家任何消息。

她麻木地往返学校与寓所，早上洗脸的时候，慨叹一具行尸还要比她活泼一点儿。

正当她以为与烈家的关系告一段落时，烈战胜却到夏宅来找她。

荷生开门进去，看见他与母亲正在闲话。

他们在谈关于移民的问题，从母亲钦佩的神情看来，烈战胜一定提供了不少忠告。

他见到荷生，立刻站起来。

这一次，荷生发现他脸上有太多的哀伤。

“荷生，我想请你去看看烈云，也许会唤起她若干记忆。”

荷生点点头。

一路上，烈战胜没有再说话。

烈云已经返回琪园。

她穿着整齐，坐在安乐椅上，看到荷生进去，一脸笑容。

荷生伸出手臂，“烈云，你认得我，说你认得我。”不由自主，泪流满面。

烈云见她哭，吓一跳，踌躇起来，收敛了笑容，狐疑地看着荷生。

不，她没有把她认出来，她似受惊小兔似的瑟缩在椅中。

荷生过去抚摸她的脸，“烈云，烈云。”

烈战胜在旁边一声不响。

看护过来干涉，“小姐，请勿影响病人情绪。”

荷生只得神色呆滞地退出房间。

良久她才抬起头问："烈火呢？"

烈战胜答："我让言诺陪他出去散心，暂时他不宜留在本市。"

"你要不要我陪着烈云？"

"你能每天来与她聊天就好。"

"我愿意。"

"司机会来接你。"

"烈先生。"

他转过头来。

"我能不能问你几个问题？"

他看着她，从来没有人敢这样问他，他亦从来没有机会说过一句心中话。

她比他第一次见她时瘦得多，也憔悴得可怜，一个无辜的外人，为了烈家的缘故，受尽精神折磨。烈战胜低声答："你当然可以，请随我来。"

荷生跟他走到楼上，他推开一扇门，里边是一个宽大的私人书室，长窗对着花园。

荷生走到窗前往下看，她的记忆告诉她，有一次，在参观花园的时候，她发觉有人在露台看她，"是你。"她冲口而出。

烈战胜正在斟酒，"是，"他答，"是我。"

那天，他听到银铃似的笑声自窗缝钻进来，他遭了迷惑，谁，是谁有这样的笑声？他已有多年未曾笑过，也有多年未曾听过如此可爱悦耳的笑声。

他放下手中的文件，不由自主地走到窗前俯视。

他看到的是夏荷生。

可惜夏荷生现在也不大发出那样的笑声了。

荷生坐下来。

“你的问题呢?”烈战胜像是已经准备好。

荷生抬起头，“琪园原本属于周琪女士，可是真的?”

“屋子的确由她父亲所建。”

“现在你是它合法业主?”

“是。”

“你如何得到它，你又如何承继了周氏大部分产业?”

烈氏不假思索地答:“一切由我合法赚得。”

“怎么样合法?”

“很简单，即使你也听得懂。十三年前，周氏被控涉嫌行骗，而实际主使人是周琪与银行主持朱某。周氏在案发前一直被蒙在鼓中，兵败后由我与言氏通过私人及业务上的关系，得到六家公司援助，注入资金，令烈氏不致倒闭，琪园早已成为抵押品，其后由我本人赎回，此事路人皆知。”

“周琪背叛她的父亲?”荷生觉得难以置信。

烈战胜看着她，“看样子你情愿相信烈战胜强行霸占周氏产业。”

荷生深深吸进一口气。

"还有没有问题？我怕你受不了这些答案。"

"有，"荷生固执地说，"还有问题。"

烈战胜叹口气，再斟一杯酒。

"烈风是不是你的孩子？"

烈战胜讶异地转过头来。

荷生自他眼神上得到答案。

"不，他姓朱。"

"啊！"

"现在你明白了。"

如果这一切都是真的，那么，烈战胜不是坏人。

荷生忽然歇斯底里地笑，抑或只是面部肌肉不由自主地抽搐？天底下哪里这么容易分黑与白、忠与奸、好与坏。

她伸手，抚着面孔，才收敛了这副悲惨的笑脸。

"一时接受不来吧？"

荷生不知如何回答。

他说下去："周氏是我恩师，当年由他恳求让这个外孙姓烈，我没有拒绝。"

书房完全静寂下来。

过许久许久荷生才问："一家人怎么会有那么多的恨？"

烈战胜看着她，"你还愿意成为这个家的一分子吗？"

“为什么把这一切都告诉我？”

他简单地答：“因为你问我。”

这当然不是真实答案。

他放下酒杯，转身离开书室。

荷生一个人坐在房内，情绪激动。

她已听过周氏与烈氏的故事，如果言诺肯把他的版本也告诉她，当年的恩怨，就会变得更加立体。

回到家中，荷生发现母亲已收拾好行李。

她缓缓坐下，惘然地想：要独立生活了。她曾经向往过这种自由，但它一旦真正来临，她又满心不是滋味。

夏太太出来看见她，“荷生，那位烈小姐情况如何？”

“烈先生已聘了良医。”

夏太太似有点放心，“如今没有医不好的病。”

心病呢，心病又如何？

“烈先生十分热心，给我几个联络人的地址，相信有用。”

“你几时动身？”

夏太太一呆，“荷生，我早说过好几次，是下个星期一。”

哎呀，荷生呆呆地看着母亲，她一次都没有听进耳朵里，为了使母亲放心，她强笑说：“我故意不要记得。”

“你随时可以来，这并非生离死别。”

“你也是，假如移民生涯不适合你，马上回头，切莫犹疑。”

“当然，”夏太太笑，“我可没有包袱，我可无需争一口气给什么人看。”

荷生握住母亲的手。

送走母亲那日，荷生才发觉她还没有换季。

自飞机场返回家中，她收拾东西，却找不到最好的两件开司米毛衣，便扬声叫“妈妈——”出了口才想起母亲正飞越大西洋，寂寥地坐下。

小小公寓似有回音。

门铃骤响，荷生去开门，言诺站在门口。

他说：“我竟没来得及去送行。”

荷生庆幸她刚洗过头化过妆，看上去不致太过憔悴，她衷心欢迎言诺。

他已穿上了灯芯绒西装，可见天气已经凉快。

“听说你通过了？”

荷生点点头，讲师们有心放她一马。

“你刚回来？”

言诺答：“昨天。”

“烈火好吗？”

“你们两人到底怎么样？”

“我不认为他会原谅我。”

言诺脱下外套，搭在椅背上，“他最近情绪不稳定。”

荷生苦笑。

言诺忽然问："荷生，你们在一起，到底有没有快乐过?"

荷生十分尴尬，"我无意把私事摊开来说。"

言诺不以为然，"你我之间，还有什么话是不能说的?"

荷生吁出一口气。

"烈火把胡子又留起来了。"

荷生低下头。

"烈云这两天有进步，认得熟人，但叫不出名字。"

"这是好现象。"

"看护说你这两日没去。"

"我在家陪母亲。"

"现在可有空儿?"

荷生点点头。

烈云的睡房里摆满了医疗仪器，设备与最先进的病房差不多。

她在看书。

见到荷生，她侧着头想一想，"你好久没有来了。"

荷生趋向前去，"你知道我是谁?"

烈云笑，摇摇头。

看护温和地说："痊愈需要时间。"

荷生抬起头，"也许她不想再有记忆。"

看护一怔，“哦？”

荷生低声说：“如果有选择的话，我亦愿意把若干记忆片断清洗。”

看护微笑，“事情不至于这么坏吧。”

荷生苦笑。

她拾起烈云在看的书，“《快乐王子》，噫，我最喜欢的故事之一，”她问烈云，“我读给你听好不好？”

烈云指着图片，“燕子。”

“是的，”荷生很高兴，“这是快乐王子的燕子，你看，结果它没有南飞，为了帮助别人，它死在王子铜像的脚下。”

说到这里，荷生皱了皱眉头，童话故事的结局往往出人意料，且残酷地写实，十分悲凉。

“但是天使来接它回去，看。”烈云叫荷生看图。

这倒是真的。

荷生握着烈云的手，“多么好，你已会看故事书了。”

烈云也笑。

荷生把她楼在怀中，烈云像一个三四岁的孩子，呵，这真是人生的黄金时代，对烈云来说，未必有什么损失。

言诺敲门进来。

他轻轻问：“你觉得小云怎么样？”

“我认识她这么久，觉得她最开心的是现在。”

“荷生，你不应这样说。”

“言诺，你看着烈云长大，你比我更清楚，生在一个这样的家庭里，爱着一个彻头彻尾利用她的人，一直做着明争暗斗的核心，你说，有没有幸福?”

“我们都希望她会痊愈。”

“当然。”

看护说：“我要与烈小姐到花园散步。”

荷生站起来，“我们走吧。”

荷生知道事情不会从此结束。

有人要偿还这笔债。

来到大门口，言诺把车驶过来接她。

荷生眼尖，一眼看到树丛另一头停着一部车子。

她认得它。

忽然之间，荷生无法控制情绪，她一伸手，把言诺推下车子。言诺冷不防她这样力大无穷的一推，踉跄落地，一边大叫：“荷生，这是干什么?”

只见荷生抢上驾驶位，右脚用力踩下油门，车子飞驰出去。

言诺插手，“荷生，停下来，危险，停下来!”

夏荷生充耳不闻，直向树丛那角撞过去。

那一部车子的司机眼见小轿车迎面扑来，慌忙间完全不能做出反应，说时迟那时快，轰然一声巨响，车头已经吃了一记，撞得对方车头灯粉碎。

荷生身子向前一冲，她随即转排挡，车子往后退，看样子她打算再来一次。

言诺惊呆了。

树丛下边就是悬崖，下去三十米左右是一条公路的回环天桥，夏荷生到底想怎么样？

只见她再次向那部跑车迎头撞去，那司机怕了，跳下车来，往私家路直奔。

言诺看清楚了那人的身形面孔，忽而镇静下来，不再出声呵斥荷生，他紧紧握着拳头。

夏荷生见逼出司机，丝毫不放松，转舵，直追，车子落斜坡的速度惊人，眼看追上那人，要朝他身子撞过去，那人惊恐之余，摔倒在地，一只葫芦似的滚下山坡，荷生并不放松，直驶到他身边，才“吱”地一声踏定刹车。

她下车来。

那人是烈风。

他已经摔破了额角、膝头，衣服上也有撕破的痕迹。

夏荷生指着他说：“滚，滚！”

他爬起来，双眼盯着荷生，荷生吓一跳，这是人的眼睛？不不，怎么两目通红如一只怪兽。

荷生鼓起勇气踏前一步，“不要再出现在这块私家地上，否则不要怪爹娘生少你两条腿。”

这时烈风忽然仰起头怪笑起来，“你们一家正在受苦是不是？我就是要你们受苦！”

言诺这时候奔到荷生身边，拉住她。

烈风一跛一跛地走下山去。

荷生激动得浑身颤抖。

过很久很久，言诺忽然说：“我不知道你会开车。”

荷生神经质地笑起来，随即颓然地坐在路边石栏上喘气。

这时琪园里的人都出来看发生什么事。

言诺的小轿车头部团成一堆，荷生这才晓得怕。

“我们也并不能肯定那是他。”

荷生坚持，“是他，是他设计诱烈云出走，是他策划由我做中间人，嫁祸于我。我误信他有苦衷，害了烈云。”

言诺扶住荷生的肩膀。

他忽然说：“伯母临走之前与我通过一个电话。”

“什么？”

“夏伯母叫我照顾你。”

荷生叹口气，母亲说得对，她的确不能照顾自己。

“来，我送你回家。”

他到车房去开出一辆烈家不常用的小车子，载荷生走。

从头到尾，烈战胜父子并没有出现，他俩不在琪园。

途中荷生问："烈风来干什么？"

"他说得很清楚，他来看我们吃苦。"

"精神有毛病的是他，不是烈云。"

言诺说："如我说，烈家的事，十分复杂。"

"言诺，烈风姓朱，不是烈火的大哥。"

言诺不语。

"这是烈战胜亲口同我说的。"

言诺专心驾驶。

荷生觉得事有跷蹊，"你知道什么？言诺，告诉我。"

"我只知道烈火是我好友。"

荷生轻摇头，"言诺，你真是一个有美德的人。"

言诺笑笑。

"令堂仍然生我的气？"

"好多了。"

"有没有为你介绍女伴？"

"你关心吗？"

"是，我关心，只有十全十美的女孩子才可与你匹配。"

"但我配不上你。"

"你现在已经看清楚我了。"荷生苦笑。

"三分秀气，四分傻气，加三分运气，我不担心你。”

“十分受气。”

言诺只是笑。

“要不要到我家来喝咖啡？”

言诺停好车，心头有点酸，把头伏在驾驶盘上，轻轻问：“怎么见得我是一个没有血性的好人，叫我来就来，叫我去就去？”

荷生张大了嘴，“对不起，你误会了，我没有非分之想，我只是请老朋友喝杯咖啡，我不是那个意思。”

她推开车门，匆匆上楼。

荷生只觉头晕身热，双耳烧得热辣辣的。

荷生太后悔说得那么多。

事情怎么可能同以前一样？

她低着头自手袋中掏钥匙，忽然之间，有一只手伸过来搭在她肩上。

荷生猛地转过身子，只看到一把长头发与一脸的胡须。

“烈火。”

他紧紧地拥抱她，在她耳畔说：“送你回来的那家伙若不是言诺，我会叫他好看的。”

荷生说：“暴力一定要停止。”

烈火看着她，“啊，听听这话出自谁口，刚才有目击证人同我说，有一辆车子意外失控，不料撞上另一辆停在崖边的跑车，跟着自动溜下斜坡，相信是手刹出了毛病。这些，都不算是暴力？”

荷生苦笑。

她掏出钥匙，开门让烈火进屋去。

“言诺说伯母外游去了？”

荷生点点头。

烈火坐在安乐椅里，“荷生，我也想过，如果我要将你托付给一个人，最理想的人选也是言诺。”

荷生紧绷着脸，“又不是包裹，何用托来托去，你要是不满意现状，干脆一声再见就行。”

烈火挨了骂，也不出声。

过一会儿他说：“伯母不在家，我倒反而规矩起来了。”

以前他总在走廊里拉住荷生，希望多聚一刻。

不为什么，只为不甘心，待听到夏太太咳嗽，才肯开门离开夏宅。

现在他忽然怀念这一声假咳嗽，此刻他不知道什么时候走才好。

他知道伯母不喜欢他。

伯母希望荷生毕业后到中学任教，嫁给言诺，自此过平淡安乐的日子。

烈火笑了，喃喃地问：“没有后悔？”

荷生诧异地问：“你说什么？”

烈火打一个哈欠，“你这里好静好舒服。”

“难怪我一睡可以一整天，你累了就休息一会儿。”

烈火索性将腿一搁，打起瞌睡来，多日来发生的事令他筋疲力尽。

此刻他努力要睁开双眼，竟不能够，轻轻叹出一口气，任由灵魂进入睡乡，俗世一切，渐渐淡出，感觉舒畅无比。

荷生却不倦。

奇怪，躲在自己家中，烈火又在身旁，照说应当十分安全，为什么那种被偷窥的感觉又来了。

她轻轻走到临街的窗前，拉开一点点窗帘，往下看，却一点异象都没有。

太敏感了。

日静无事，心头渐渐空灵，听到各种几乎不存在的杂声，荷生警惕，可别看见什么怪事才好。

她想像这一切都已过去，她与烈火，终于在一起生活，烈云在周末来看他的，对平凡的假日嗤之以鼻，而言诺却说："噫，荷生，早知你要求这么低，我也可以做得到。"随即他与他美丽而贤惠的妻赶去参加一个重要的宴会。而烈火悻悻地说："看，人家取笑我们的幸福，怎么办？怎么办？"

荷生喜欢中年，一切可能性已发展殆尽，只剩下铁定事实，大多数困难早已克服，所以中年是安定逸乐的，受挫折也懂得应付，荷生盼望中年速速来临，丢掉彷徨。

烈火睡熟，面孔出奇安详，他不似言诺，表情异常丰富，七情六欲，喜怒哀乐，统统露在脸上，荷生知道这种人吃亏，却爱莫能助，心中无限怜惜。

烈火的手垂在椅旁，荷生想去握住他，又怕吵醒他，自从认识烈火以来，这是最安静可贵的一刻。

电话铃骤响的时候，荷生不知多后悔没把插头拔出。

她连忙把它拿到房间去听。

是母亲，抱怨女儿经常不在家，继而叮嘱生活细节，荷生唯唯诺诺，待母亲教训完毕，回到客厅，只见烈火已醒。

他温柔地看着荷生，“听你的口气就知你在敷衍伯母。”

荷生蹲下来，“将来我不要生女儿，她们太不像话，完全向着陌生人。”

“你不该让我睡着，此刻有点头昏脑涨。”

荷生斟了一大杯冰水给他。

醒了，一切烦恼各归各位，点一次名，一个不少，全体似一只怪兽蹭蹲在原位虎视眈眈，烈火叹一口气，把冰水杯子印在太阳穴。

他不想醒来。

电光石火间，烈火想起小云，她也许就是永远不想再度苏醒的最佳例子。

烈火叹息一声。

“做你还有牢骚，做我们更加不得了了。”荷生只得这样安慰他。

“谁会想做我?”烈火拉着她的手。

“问题能够一一解决。”

“你真乐观。”

“倘若不会消失，也只得学习与之一同生活。”

“像肿瘤一样，姓烈的人先天性一生下来体内便长着这种东西。”烈火按一按胸膛。

“事情没有这样坏，烈云慢慢能够痊愈，我们可以忘却整件事。”

“有人不惜一切工本来要我们吃苦。”

“那我们更加不能放弃。”

“那么让我俩结婚吧，举行最盛大豪华的婚礼，帖子发到每个敌人与朋友的手上，你说好不好?”

“我以为只有小女孩子把婚姻当做逃避现实的手段。”

烈火笑了。

“我要走了，”他看看表，“公司等我开会。”

荷生拉他起来。

他说：“我可以在这里睡上一辈子永不醒来。”

荷生连忙说：“三天三夜已经足够。”

“荷生，你随时可以搬到琪园来住。”

荷生对琪园没有一丝好感，只是微笑地说：“你想让我代你照顾药园?”

“这是其中一个原因。”

“还有什么理由?”

"我可以天天看到你。"

荷生黯然："有一度我还以为你不要再见我呢。"

"对不起，荷生。"他把脸埋在她手中。

荷生送他出门。

"考虑一下，到琪园来住。"

荷生不想使他失望，只得推搪说："让我想一想。"

烈火走了。

屋内掉一根钉子也可以听得见。

电话铃声又响起来。

荷生以为是烈火，连忙取过话筒。

"夏荷生，夏荷生。"

荷生一听到这个声音，如闻招魂，急想挂断，但随即明白此举太过助长对方威风，便尽力控制情绪，"我是夏荷生。"

"既往不咎，请告诉我烈云现况如何。"

荷生恨不得捏死这个人，嘴里却说："我劝你马上挂断电话，以后都不用企图与我联络，否则我会交给警方处理。"

她的声音十分坚决，务必要对方得到正确信息。

对方却缠上来，"告诉我烈云的近况，我答应你马上挂断——"

荷生忍无可忍，把电话插头拉掉。

他分明是欺她一人，或因她落单，或因她心软，他一直看穿这点，咬牢她不放。

搬到琪园，或许可以避开此人骚扰。

深夜，有人敲门，荷生胆战。

门外却是言诺，“你为何不听电话？”

荷生答非所问：“言诺，我们为什么不把他交给警方？”

言诺沉默。

他完全明白荷生说的是什么。

“那人骚扰你？”

“不要再用私刑报私仇了，言诺，同烈先生商量一下，交给警方处理。”

“烈先生不会那样做，其中牵涉太广，审讯起来，证供会毁了烈氏一家。”

“你想那个人会不会罢休？”

“荷生，那么你暂且来我家住。”

“你家？”荷生失笑。

言伯母大概等着奖她一巴掌呢。

“不不不不不。”

“荷生——”

荷生恳求地低声说：“不。”

“那么，搬到琪园去。”

“一个人没有自己的家，实是非常悲哀的一件事。”

“独居人要分外当心。”

言诺自公事包取出一只手提电话，“我要你用它。”

荷生点点头。

“晚上要出去，我这边有车。”

“谢谢你。”

“荷生，你太倔强了。”

荷生微笑，“你也是。”

“加上烈火，一共三个。”言诺苦笑。

现在只余烈云最温驯，但是每一个人都希望她速速恢复原状，做回那个任性不羁、生活在一人浪漫世界里的小云。多么讽刺。

“我要走了，烈火在等我。”

荷生讶异说：“你们俩真的深深爱着对方，现在我相信了。”

言诺指着荷生，“这件事要弄清楚，我并没有把你让给烈火，是你主动舍我而去的。”

“尽管责怪我好了，自古祸水还真的都是美女。”

言诺打开门，“我一走你便上锁。”他不想与她斗嘴。

他去后，荷生并没有依嘱锁门睡觉。她换过衣服，约好同学，外出聚会。

这些日子，处处以烈家的人烈家的事为中心，几乎忘记自己是谁。

同学的车子停在门口，荷生在上车之前看到一弯蓝月，她牵牵嘴角，登车而去。

同学说："荷生，好久没有与我们出来，听说你心情欠佳。"

荷生看他一眼，"何止心情，名誉大概也差不多水准。"

两位男同学都笑，"名誉倒不值一哂。"

"大学生说出这种话来，叫人心寒。"

"大学生一毛钱一打，叫我们说得出什么好话。"

荷生许久没有这样毫无心机说说笑笑，无聊有趣，觉得十分享受。

"听说两位男士令你不知取舍，烦恼得要死。"

"不就是你们两位吗？"荷生也很会调笑。

同学吐吐舌头，"我们可不打算为女生打破头。"

言诺亦没有这种打算。

"我们还听说有第三个第四个。"

荷生一怔，啊，真的传得那么厉害？流言可畏。

"对呀，都等着老校长传你进去训话，勒令退学。"

荷生见他俩语气愉快到不堪的地步，便悻悻然说："校长问起，我就报上贤昆仲的大名。"

大家都笑。

"真的，荷生，都传得不像话了，或许你该收敛点儿。”

荷生无奈地答：“事情完全不是这样的。”

“言诺是品学兼优的小生。”

“我知道。”

“你又何必同那家人的父子兄弟搞在一起，据说连母亲都气走了。”

“什么，”荷生拉下脸，“再说一遍。”

两位同学交换眼色，连忙噤声。

“再说一次。”

他们不敢再提。

“停下车来。”

“荷生，大家老同学了——”

“我不认识你们，你们也不认识我，没有必要同车而行。”

“荷生，对不起，他们说错了，大伙在等我们，别节外生枝。”

“他们说错，你为什么不更正他们，明知是错，还把话在我耳畔重复一次，叫我难堪，你比他们还坏，他们并没有认是我的朋友。”

同学也是年轻人，也气上心头，把车停在一边，“夏荷生，你对朋友太苛求太计较了。”

荷生推开车门，“我为什么要故作大方同你们虚与委蛇？我不必降低要求，我不要这种朋友。”

下了车，凉风一吹，人一清醒，夏荷生不禁失笑。

不要这种朋友，恐怕永远交不到朋友了。

抬头一看，人家的车子并没有开走，慢慢吊在她身后，看她会不会回心转意，这样的朋友，已经非常难得。荷生扬起手，车子停下来，她再度上车。

荷生决定继续玩这个游戏。

同学轻轻说："朋友呢，不过是互相协助对方杀死时间的帮手，太认真就不好玩了。"

荷生大声说："说得好说得妙。"她大力鼓掌。

许多喝酒的地方不招待单身女客，荷生需要他们带路，否则孤掌难鸣。

一共十来个同学坐一张台子，闹哄哄，浑忘烦恼。

酒过三巡，荷生觉得宾至如归，在嘈杂的乐声中与同学们搭着肩膀起舞。

午夜时分，大家也就散伙。

男孩子们细心地把荷生送回家，且陪到大门口，看她用钥匙启门进屋，才返回车上。

荷生站在露台上向他们招手。

在旁人眼中，他们全是小阿飞吧。

荷生退回屋内，关掉露台灯，半夜，只她这一间屋子有一线亮光，太惹人注目。

荷生拥着被褥睡着。

听到门铃响时，天色尚未大亮，荷生一时醒不过来，终于挣扎着坐起，也要着实过一会儿，才能肯定身在何处。

她蹒跚走到门前，在防盗孔张望，没提防看到一双血红的眼睛。

荷生退后两步，取过电话，拨了两个九字，又放下。

她终于打给言诺。

“荷生，什么事?”

“他在我门口不停地按铃，我想报警是最好办法。”

“坐下，保持冷静，我马上来。”

“我给你十分钟，他要是继续胡闹，我立刻报警。”

门外传来吼叫声，“夏荷生，我知道你在里边，我与你讲几句话就走。”

荷生问言诺，“你听到没有?”

“我现在马上出门。”

荷生挂上电话。

那人在门口号叫：“告诉我烈云怎么样?”

荷生忍无可忍，拉开大门，隔着一道铁闸，与他对质：“烈云怎么样，你有一百个方法可以打听到，何用到这里来耍赖!”

他看到荷生出来，语气转为哀求，“跟我说几句话。”

荷生说：“你有病，你一直有病，你要去看医生。”

“我不知道这件事，烈云遭绑架与我无关，你要相信我，当时我不在本市。”

荷生听到整齐的脚步声传来。

是警察，邻居不胜其扰，向派出所报告。

果然，梯间转出两名制服人员。

“什么事?”他们抢上前来。

接着，言诺出现了。

荷生打开门。

警察说：“小姐，这里是住宅区，不容你扰攘，有什么事，最好静静解决。”

两男一女，还不是三角恋爱纠纷?

荷生看着言诺，言诺对警察说：“我们会和平解决的，麻烦诸位了。”

“肯定不用协助?”警察问。

“是的。”

警察查过他们的身份证明文件后离去。

言诺这个时候才转过去对牢烈风，“有什么事，你还是直接对烈先生说的好。”

烈风呆呆地看着他俩，过一会儿颤抖着声音问：“我能不能见烈云?”

“这根本不是夏荷生可以办得到的事，你何必来骚扰她。”

烈风的神智似乎恢复过来，他呆了半晌，自梯间下去。

他会再来。

他看得出整幢墙最弱的关节便是夏荷生，必须自她这里入手。

荷生返回屋内，经过这一段事，她累得倒在沙发上。

她跟言诺说："他说他是清白的。"

言诺答："人一照镜子，必然看到最清纯洁白无辜可爱的影像。"

荷生苦笑，"难怪我浴室那面镜子要爆裂。"

"你的推测是什么样的?"

"很悲观。"

"说来听听。"

"言诺，我以为你闲谈绝不说人非。"荷生意外。

言诺摆摆手。

荷生开始推测，"那日清晨七时，烈云离开这里，便出去与那一帮人会合，接着他们把事前准备好的信送到琪园。本来，烈云打算提出要求，盼望烈先生恢复烈风的地位，但是，人家发觉到这是一个千载难逢的发财的好机会，结果烈先生必须付出赎款。"

"烈风扮演什么角色?"

"他是主谋，策划一切，然后到外国去躲了几天，原来最简单不过的一个游戏失却控制，他也料不到烈云会受到极大的惊吓以致失常。"

言诺沉默一会儿，"你认为弄假成真是一桩意外?”

荷生一怔，“什么，你说什么?”

这问题一脱口，荷生想到了一个人，她看着言诺。

言诺说:“你猜到了。”

“当然，”荷生惊道，“是周琪女士。”

言诺不出声。

“她恨烈战胜已到极点，天。”

言诺低下头。

“他们决意一生要叫对方受苦。”

言诺叹一口气。

荷生不能置信，“烈风与烈云他们成了武器与炮灰。”

言诺站起来，“今早有没有课，我送你回学校。”

“言诺，你好镇静。”

“我们不过在编故事，也许真实情节并非如此，我们不能肯定。”

荷生呆了半晌，她紧握着拳头，松开手的时候，指甲已掐进掌心，印起红痕。

他们并没有把这件事告诉烈火。

荷生隐隐觉得这是一个不可弥补的错误。

为了避免应付烈火过激的反应，渐渐她会瞒他更多。

那天下午，烈火找到荷生，跟她说，稍迟来接她去琪园一聚。

荷生急着更衣，不知怎地，翻遍了衣橱，都找不到稍微鲜色的衣裳。

黑白灰流行得太久了。

稍早时她似乎看到女装店挂出骆驼色的毛衣裤，可惜无心置装，错过机会。

去年一套天蓝色衣裙又无论如何找不到，荷生看着一大堆不能搭配的衣服，深觉命运弄人，此乃缩影，索性把它们推入樟脑箱内，推上盖，眼不见为净。

她只得挑件奶白色宽身裙套上。

裙子近胸有一朵织出来的玫瑰花，铁锈色，夏太太见了曾皱眉道："这是什么，似一团番茄酱，又似干涸的血渍。"

荷生当时觉得人生最大的荆棘便是要讨好老妈，但今日，她照着镜子，发觉老太太的联想并非空穴来风。

已经没有时间了。

烈火已在按铃。

荷生分外不安。

到达琪园，连忙要一口酒喝。

言诺早在等他们，令荷生意外的是，烈云端端正正坐在图画室，打扮得非常整齐，一如平时。

荷生迎上去，"烈云，你气色好多了。"

烈云笑，大眼睛空洞地看着荷生，仍然没有记忆。荷生坐在她旁边。

烈战胜自花园进来，“荷生，真高兴看见你。”

荷生抬起头，他两鬓似添了白发。

荷生勉强笑道：“今天是什么大日子？”

“没有事，很久没有在家吃饭了。”

荷生握着烈云的手，烈云把头靠在荷生的肩膀上。

烈战胜看到这种情形，告诉荷生：“烈云的母亲要把她接走。”

荷生意外，“可是烈云要接受治疗。”

“女人不可理喻。”

荷生只得道：“医生或许可以说服她。”

烈火进来，“父亲，今天有事商议？”

“我们庆祝雨过天晴。”

荷生呆住了，乌云密布，何来一角青天？

她低下头，不予置评。

言诺本来最习惯烈家作风，但这次他也露出不自然的样子来。

烈云不知听懂了哪句话，忽然轻脆地拍起掌来。

荷生连忙再喝一口酒。

不知怎地，烈云好端端又哭起来，伏在荷生身上饮泣。

言诺找来看护，把烈云送回房间休息。

烈战胜沉默了。

荷生觉得她有责任顾左右而言他，因此无稽地说："学生生活真不好过，很多时候都想辍学。"

言诺说："中学与大学之间，最好留一个空当，体验一下生活。"

就在这个时间，荷生听到花盆碎裂声，她抬起头来，荷生的耳朵最灵，她发觉室内其余三人没有注意。

莫非是多心了。

荷生又低下头。

言诺说："烈先生，反正还有时间，不如谈谈公事。"

荷生点点头，"真的，不谈公事，仿佛无事可谈。"

她站起来，"我出去走走。"

荷生走到花园，心底那股不安，渐渐上升，她兜回走廊，上楼去看烈云。

看护在会客室听音乐翻阅杂志。

荷生走到房外，浑身汗毛忽然竖立，她轻轻推开房门，看到烈风蹲在烈云跟前。

他必有琪园整套钥匙，一定由烈云私授予他。

荷生连忙掩上门，"快点走，趁没有人知道快点走。"

烈风受了刺激，看到荷生不避开反而迎上去，"她不认识我，烈云不认识我。"

说着他泪流满面。

烈云自言自语道："父亲会骂，二哥哥也会不高兴。"

"烈风，我要你马上走。"

烈风恼怒，"你是谁？你也来喝令我？"

荷生过去拉他，"你自露台进来是不是，快走。"

烈风一手把荷生推开，"我还以为你与他们不同呢。"

荷生无暇分辩，趋向前去，要进一步推走他，但是烈风已经红了双眼，他一手拉起烈云，另一手甩开荷生，荷生左脚在地毯上一滑，撞向台角，额头一阵剧痛，但是奋力扑向前抱住烈云双腿。

烈风猛然取过椅子，向荷生撞去。

荷生只觉得面孔上滑腻腻，不知道已经血披满面，她吓怕了烈云，烈云大声哭泣。

说时迟那时快，门外一声吼，烈火扑进来，抓住烈风，荷生刚刚扶着言诺的手爬起，只看见他们两人卷向露台，撞碎玻璃窗，其中一人似鹞子似的飞出栏杆。奇怪，荷生觉得一刹那天地间失却音响，一切停顿，但荷生清晰地看到烈风衣裤飘飘，堕下楼去。

继而听到巨物坠地声，"轰隆"一下，众人尖叫起来。

荷生推开言诺，跌撞着抢到露台，烈火一身血渍斑斑，手臂上还插着碎玻璃。

荷生伸出手去，“烈火，烈火。”

烈火转过头来，. 很平静地看着荷生，过一会儿，他伸出手来，把荷生湿透的碎发拨向耳后，紧紧拥抱她。

荷生把头贴在烈火胸前，不肯放手。

耳畔杂声纷沓而至，救护车与警车号角，制服人员的脚步声。

终于有人拉开荷生，荷生的额角犹如开了洞，血汩汩流出，她却一直能够保持清醒，她缓缓走到楼下，在走廊的深色镜子内照到自己，浅色裙子上一搭一搭全是拳头大血印玫瑰花，她忽然明白夏荷生已经偿还花债，一颗心遂静下来，再也没有一贯忐忑不安的感觉。

救护人员替她按住伤口，荷生转过头来，看到言诺惊恐莫名的表情，他牙关打战，人抖得犹似一片落叶，他害怕，平素镇定的言诺怕得脸色如一页白纸。

经过花园，荷生看到烈风被抬上担架。

他四肢犹如提线木偶，折向不可能不合理的方向，荷生看到他凝固的眼珠，那股仇恨的鲜红色已经褪去。

整间琪园为之沸腾。

荷生踏上救护车。

一躺下来，她看到车顶那盏灯逐渐模糊，淡出，四周围变成漆黑一片。

醒来的时候，荷生躺在医院的病床上。

她首先看到烈战胜憔悴焦急的面孔。

“荷生。”他握住她的手。

荷生在这一刹那清醒过来，前尘往事统统归位，她虚弱地问：“烈火，烈火。”

烈战胜把嘴凑到荷生耳边，“他平安。”

“烈风怎么样？”

烈战胜急促道：“荷生，他已过世。”

荷生闭上双目。

烈战胜身后的制服人员上前说：“夏小姐，你可否回答我们几个问题？”

医生看看表，“十分钟，你们统统给我出去。”

警务人员问：“昨夜，琪园二楼的睡房中，发生了什么事？”

荷生转一转头，剧痛使她露出苦楚的表情。

过一会儿她说：“我推开房门，即受袭击，接着有人跳楼。”

“他自己跳下去的？”

“是。”

警员凝视她一会儿。

荷生无惧地回望，他是一个端正深沉的年轻人。

“没有打斗？”

“他殴打我，旋即把我们推开，撞破玻璃跃下。”

“我们?”

“烈火试图抓住他，但不成功。”

“你看到的只有这么多。”

“是。”

警员站起来。

烈战胜走过来，“荷生，你请休息。”

他随警员退出。

荷生庆幸母亲不在本市。

她独自呆视天花板直到下午。

言诺来看她，两人恍如隔世，半晌不知道该说什么话。

终于她问：“烈云怎么样?”

“案子一结束，她母亲便接她到英国疗养。”

“案子，什么案子?”

“荷生，烈火被控误杀。”

荷生不出声。

她别转脸去。

审讯期间，烈火未能获准保释。

荷生去探访他。

烈火只肯见她一次，他对她说：“你要是真肯让我放心，马上同言诺结婚，去。”

荷生当时不发一言，站起来就走。

言诺追上去，看到她泪流满面。

这些日子，也只有他陪着她。

言诺还得在长途电话中帮着安抚夏太太，背着良心说谎："完全同我与荷生无关，那只是烈家的事，伯母，你完全不用赶回来，我们天天可以与你通消息。"

荷生在法庭上始终坚持同样口供。

忽然之间，公众席上一个穿黑衣瘦长的中年女子站起来指着她骂："夏荷生，你隐瞒事实，你明知他被推下致死，你是帮凶，你永生永世不得安眠。"

荷生认得她，她是周琪。

庭内大乱，陪审员耸然动容。

周琪被请出法庭。

那天，荷生无法独处，她由言诺陪着，到新居去看烈云。

烈云不肯让她接触，像是害怕憔悴落形的荷生。

"烈云，是我，是荷生姐。"

烈云侧着头，好似对这个声音曾经熟悉。

言诺深深叹气，走到房外。

荷生正欲放弃，忽然之间，烈云抬起头来。清晰地问："他已经不在了，是不是？"

荷生呆住。

"他终于不再烦恼了。"烈云吁出一口气。

"烈云，你说什么，烈云，你是否已经痊愈。"荷生大声问她。

烈云跑到另外一个角落，护士听到异响连忙赶进来，荷生知道一切仍是幻觉。

言诺送她回家。

途中荷生说："现在谁也不能住在琪园，大屋终于废置，争无可争，不再烦恼。"

言诺无言。

烈战胜的车子停在夏宅门口，他与律师一起下来，告诉荷生，"案子明日作终结陈词。"

荷生在劲风中打了一个寒噤。

律师说："证供对他有利。"

荷生与言诺同时别转头去。

烈战胜上车离开。

言诺陪荷生回家，他俩彻夜不能成眠。

荷生站在露台上，看向幽暗的街道，不知怎地，她看到净是一双双血红的眼睛。

言诺说："我陪你下楼散步。"

荷生披上大衣。

"我担心你。"言诺说。

"我很好，"荷生苦涩地答，"不用做事，不必上学，做一个职业证人。"

"相信你看到了烈火的情形，他似丧失了斗志。"

荷生心如刀割。

"烈先生不让你出席旁听，真是明智之举。"

荷生抬起头，"我们会不会赢？"

"荷生，那么多人见过他们兄弟吵嘴、打架，以及烈火保证要把烈风干掉的誓言。”

“但是，”荷生拉住言诺的手臂，“我同你的证供……”

言诺无奈地说：“我同你是烈火的什么人，大家都知道。”

“你不太乐观。”

“我一向是个以事论事的人。”

荷生抬起头，看着月亮。

“记得第一次带我去琪园吗？”她问。

“怎么不记得，这是我一生中最大的错误。”

他们放慢了脚步，有一个小女孩牵着一条狗迎面而来。

不知怎地，荷生的目光为这条狗所吸引，只见它通体白色短鬃，体积庞大，气息咻咻，走近了，仰起头，对准荷生。

荷生猛地一怔，狗的双眼狭长鲜红，吓她一跳，再加注意，它的五官渐渐化为烈风的面孔，变成烈风的头镶在狗的身上。

荷生崩溃下来，她退后一步，尖叫起来，叫完一声又一声，无法停止，再也站不稳，蹲在地上。

女孩与狗早已离去，她却继续惨嚎，言诺只得伸出手，用力掌掴她。

荷生脸上吃痛，呆住，怔怔地看着言诺。

言诺不忍，紧紧抱住她。

荷生惊恐得一颗心似要自口腔里跃出来，魅由心生，倘若一生要背着这个阴影而活，真是生不如死。

第二天，荷生坐在家中等消息。

烈火一案已在最高法院聆讯完毕，六男一女陪审团退庭商议。

六个半小时之后，向法庭回报。

裁定烈火罪名成立，按察司判被告入狱三年。

荷生听到这个消息，耳畔有细微“嗡嗡”声，她低着头，双臂抱在胸前，默默无言。

律师还向她解释细节，她却一个字都听不进去。

荷生有点感激这“嗡嗡”声，希望它不要消失。

烈战胜走过来，荷生不由自主地把头埋到他怀里去。

没有棋子了，他们都没有棋子了，烈风已死，烈火入狱，烈云失常，这一场战争，胜利者与失败者牺牲得同样惨烈。

烈战胜一句话都没有说，带着荷生及言诺去见烈火。

烈火握着荷生的手，“答应我一件事。”

荷生不语，她知道他要说什么。

奇怪，在这种时候，他偏偏去关注微不足道的琐事。

“马上与言诺结婚，有多么远走多么远。”

荷生情绪不受控制，神经质地惨笑。

烈火急促地转向言诺，“你听见我说什么了吗?”

言诺点头，烈火似略微放心。

然后他主动地站起来说：“你们走吧。”

他们缄默地回到烈宅。

烈战胜一开口便说：“我要你们离开本市。”

荷生没听清楚，她的听觉失灵，身边像有一只不肯飞走的蜜蜂。

言诺向她重复一遍。

荷生点点头，“我正想去探访母亲。”

“言诺，你帮荷生去安排一切。”

言诺似有问题未能解决，他与烈战胜商议起来。

荷生走开去找烈云。

推开房门，只见一张空床，护士正要收拾仪器，看到荷生，见是熟人，便向她笑笑。

荷生指一指床，“人呢?”

“今晚起程去麻省医疗。”

“痊愈机会大不大?”

“相当有希望。”

荷生对这种高技巧的答复已经习惯。

人去楼空。

护士想起来：“对，她看到母亲的时候，会叫‘妈妈’，你说这是不是好消息?”

荷生“霍”地抬起头来，“真的？这正如在满天乌云中看到一丝金光。”

看护笑着指指耳朵，“我亲耳听见。”

“是，这真是至大至乐的消息。”

言诺上来找她，“荷生，烈先生有话同你说。”

荷生与烈战胜在书房中对话。

他温和地问：“你有什么打算？”

荷生简单地说：“等烈火出来。”

烈战胜说：“我想送你出去升学。”

“我不想再进学堂。”

“相信我，荷生，有点事做，时间会过得快些。”

荷生不做声。

“言诺本想陪你，但他不舍得长时间离开父母。”

“他一向是个好孩子。”荷生莞尔。

“你的耳朵怎么了？”烈战胜放低声音。

“什么？”

烈战胜叹口气，“荷生，要是你愿意，我可以协助你开始新生活。”

荷生微笑，“烈先生，我听不到你说什么。”

烈战胜摇摇头，“你这孩子。”

“孩子，还是孩子？”荷生失笑。

烈战胜说：“至少考虑我的建议。”

“烈先生，我一直在想，那天在琪园，如果不是我多事，上楼到烈云房间去，烈风会不会自动离去？悲剧是否可以避免？”

烈战胜抬起头来，“荷生，我永远不去检讨过去的事情。”

“即使是这件事？”

“即使是这件事。”

荷生低头看着双手。

“我安排你明天就走，言诺会陪你一个学期。”

“我怎样探访烈火？”

“荷生，他不要见你。”

“什么？”

“他已说得很清楚，他不要看见你，不要读你的信，也不要你等他。”

荷生沉默。

过一会儿她问：“为我好？”

“不，为他自己好。”

“我不相信。”

烈战胜说：“对不起，荷生。”

“就这样，一声对不起就把夏荷生一笔勾销了？”

“没有人可以这样对夏荷生，”烈战胜握紧她的手，“耐心一点。”

荷生只得点头。

烈战胜忽然问：“为什么烈家不能有你和言诺这样的孩子？”

荷生不相信他会问出这个问题来，这么聪明的人，竟连如此粗浅的道理都不懂。荷生讶异地说："正因为我们不是你的孩子。"

任何人在琪园这种环境长大，都会变成烈火烈云，甚或更加悲哀。

临走之前，荷生并没有见到烈火。

他不愿意见夏荷生。

几个谈得来的同学都来送行，见言诺与荷生在一起，心里颇有点宽慰：也许她转了一个圈子，又回到他身边去了，只要有人接手，过往不好听的花边新闻很快会淡出传为美谈，希望夏荷生可以得到较为理想的结局。

言诺搀着荷生上飞机，她同他笑，"我不是老太太。"

话还没说完，已经一跤跌在地上，吓得服务人员争相扶持，荷生挣扎着拾起手袋，一不小心，袋中物件落出来，又得一件件拣起。

荷生苦笑。

抵达西岸，她与母亲住了三天。

夏太太桌上成叠剪报，都是有关烈氏一案的新闻。

世界太小，你知道的，别人也知道，你去过的地方，别人都去过，多说无益。

从亚洲到美洲，才十来个小时的飞机，谁也甭想把谁当乡下人。

长辈脸色凝重，但看到言诺的时候，却舒一口气：荷生能够靠着这块金字招牌，就什么都不怕，一切可以从头开始。

荷生看看言诺，人们太过高估他，却低估了她。

即使如此，她也不想特地证明什么。

言诺问她："睡得好不好?"

荷生答："还可以。"

言诺有点意外。

荷生解释说："还有三年时间，没有人可以三年不睡。"

言诺明白了。

荷生与母亲道别，她不能与她住同一城市，怕会窒息，受伤的人需要额外自由与更多时间安静地来调整心理及生理。

荷生害怕每天早上起来看到母亲焦虑忧伤的面孔，迫切地、殷勤地，希望女儿在一天之内痊愈，为母亲争一口气。

荷生搬到另一个镇，租一间小小公寓，簇新的环境，截然不同的人与事，连她自己都相信可以忘记过去，从头开始。

这个镇里华人不多，没有人认识她。

荷生买到一张理想的书桌，坐下来，开始写信。

第一封信被退回来的时候，恰恰是她寄信十四天之后。

邮期很准，以后，她每寄一封信，就收到一封退信，看到信封上自己的字迹，荷生有种突兀的感觉，仿佛有一个熟得不能再熟的熟人，在天之涯海之角，找到了她，要与她通消息。

烈火不肯读她的信。

他要令她失望，死心，放弃，不收信是最直接的表示。

荷生继续写，她不是要与烈火比赛意志力，她只是想寻找一个精神寄托。

她用一格抽屉，专门来放退信。

言诺对这件事并没有发表意见，每一个人都有权对他的过去表示怀念。

在一个隆冬的晚上，言诺问荷生："有没有算过你认识烈火一共有多少日子？"

荷生想一想，讶异地答："七个月。"

才七个月。

连当事人都觉得不可思议。

过一会儿轮到荷生问："我此刻的生活费用由谁在负责？"

"我。"言诺答。

"谢谢你。"荷生一度以为是烈战胜，"你不觉得辛苦？"

"辛苦时告诉你。"

"别抱怨你动用了老婆本。"

“老婆，”言诺笑了，像是第一次听到这个新名词，“老婆。”

荷生低下头，“你已经仁至义尽，言诺，也该回去帮烈先生照顾生意了。”

“烈先生早已决定将公司逐步西迁，我们有一组人在这里部署。”

荷生意外，“言伯父也在此地？”

言诺点点头。

“呵，都把这里当行宫了。”

“烈先生做事业的心已不能与从前相比。”

荷生点点头，任凭他是金刚不坏之身，遭此巨变，怕也会灰心。

“他后天来，要是你愿意，一起去接他。”

荷生自然没有反对。

那是一个万里无云、清寒的大清早。

烈战胜看到她，即时问：“荷生，你的耳朵怎么样？”

荷生强笑着答：“一直像打着了汽车引擎似的。”

“医生怎么说？”

“没有答案。”

“我很乐观。”烈战胜拍拍她肩膀，“一定会痊愈的。”

荷生拉拉他袖子，“烈火可好？”

烈战胜声音低下去，“他没问题，可能参加一个进修计划，消磨时间。”

荷生凄酸地说：“他不肯收我的信件。”

“我已告诉过你。”

荷生动动嘴角，她总不相信他会做得到。

“他叫我带口信给你。”

“是什么，他说什么？”荷生紧张地看着烈战胜。

“他认为你与言诺原属一对。”

“叫他管他自己的事情。”荷生生气了。

烈战胜凝视她一会儿，叹口气，“有好消息给你，烈云问起你的下落。”

“太好了，言诺，过完年我们去看她。”

“别高兴太早，她的情况不甚稳定，一时记得，一时忘怀，记忆片断不能连贯。”

“但她在进步。”

烈战胜点点头，踏上来接的车子，一边对言诺说：“晚上一起吃饭。”

见面的时候，荷生却只见烈战胜一个人。

他解释：“言诺同他父亲有话要说。”

荷生一怔，父子俩有什么重要的话要说，何用千里迢迢，跑到这里来讲，思念一转，已经明白：“是因为我吗？”

“他父亲要他回去。”

荷生猜对了，微笑道：“言伯母非常不喜欢我。”她从前曾对荷生赞不绝口。

烈战胜告诉她："今天晚上他们就在这间酒店的二楼宴客，请未来亲家。"

荷生一呆。

渐渐打心底凄凉起来，当然，她不能叫言诺一辈子侍候在侧，默默耕耘，不问收获，但这么快！

她清清喉咙，"那位小姐，品貌学问都很好吧？"

烈战胜说："是老言拍档伙计的女儿。"

"言伯伯不是你的合伙人？"

"他想另起炉灶，我支持他。"

这样看来，真不能叫言诺再垫支生活费了，人家会怎么想，等那边那位小姐发话，找地洞钻都来不及，荷生知道母亲尚有一点积蓄，或许要同她商量商量。

香而甜的香槟酒在荷生口腔里变得酸涩。

烈战胜犹疑一下，把手放在荷生手背上。

荷生轻轻告诉说："言诺并没有提起他要结婚的事。"

"也许他还没有找到适当的时机。"

荷生只得点点头，静静地端起酒杯，呷一口酒。

这个时候，震中才抵达荷生心中，她明白到自己竟是一个无法自力更生的人，她渴望自由，却无能力展翅飞翔，荷生为这个事实震惊。

她推开面前的美酒佳肴，"烈先生，我觉得不大舒服。”

“我不应该告诉你。”

“不，谢谢你让我知道。”

“如果是经济上的问题——”

“不。”

“那么我送你回去。”

车还没有来，两人在门口站了一会儿，烈战胜说：“荷生，你请稍候，我去叫司机。”

荷生呆呆地看着大堂中的节目牌。

忽然之间，她听到一阵欢愉的嬉笑声，荷生抬起头来，看到三对男女迎面走来，俩老一嫩，她起码认得其中三人，他们是言氏夫妇及言诺。

只见言诺穿着礼服，彬彬有礼与女伴聊天，那女孩子肩上搭着一方轻而柔的青秋兰披肩，巧笑倩兮，容貌十分秀丽。

太不巧了，荷生自惭形秽，急急要躲到柱后。本来这种场面不难应付，大家装做看不见，便可避过，但不知怎地，言太太立定心思不肯放过夏荷生，她眼尖，立刻扬声叫：“那不是夏小姐吗？”

大家的目光自然而然落在荷生身上。

言诺只看到瘦削憔悴的她沉默地站定，像是准备接受命运的安排，但不，她的一双大眼睛里仍然闪烁着倔强的神色，嘴角虽怀凄酸，脖子却挺直。

言诺就是爱荷生这一点。

他撇下女伴趋向前去，“原来你与烈先生也在这里吃饭。”

言太太看见儿子的态度仍然如此亲昵，不禁心头有气，竟转头对丈夫说：“把别人害得家破人亡，也该知足了，莫又出来寻替身才好。”

荷生怔住，她凝视言太太。

那中年妇女已被丈夫以目光及手势阻止，颇觉得自己失仪，一抬头，与荷生的眼神接触，不禁激灵打一个冷颤，这双眼简直有毒，如一头兽般透出精光，她连忙借故走开。

荷生一生中从没被人如此侮辱过，握紧拳头，全身发热，一句话都说不出来。

言诺羞愧地向荷生道歉，“看在我分上原谅她。”

过一会儿荷生才说：“他们在等你，你还不过去?”

“荷生。”

“去吧。”

那个俏丽的女孩子折回向言诺招手，他只得归队。

言诺不满地说：“母亲，你原不必那样。”

言老却顾左右而言他，继续适才的话题。

言诺转头，看到荷生独自站在那里，身形寂寞彷徨，言诺心头一阵酸痛，忍无可忍，撇下双亲，撇下女伴与她的父母，不顾一切，大踏步地又走回荷生身边。

言太太的眼睛瞪得如铜铃，但一点办法都没有。

言诺走到荷生身边说："我送你回去。"

荷生刚抬起头，烈战胜的声音自身后响起，"怎么，我才离开五分钟，好像已经发生了许多事？"

荷生如遇救星，"烈先生，你回来了？"

"车子马上到。"

言诺低下头，对于未能及时保护荷生，惭愧不已。

烈战胜一出现就控制了场面，那班人如小学生见到训导主任，个个循规蹈矩起来。

烈战胜与他们招呼过，才与荷生上车。

他讪笑道："真不应该离开你。"

荷生面对着车窗不语。

"让我告诉你一个故事。"

荷生喜欢听烈战胜说故事，他的表达能力强，故事情节又丰富，荷生但愿他时常有说故事的兴致。

"我小时候，住在继园台附近。"

荷生不以为奇，该区在五十年代新移民最多。

"一日放学无聊，在附近溜达，竟在山间发现一座秋千架，大乐，偷偷玩了一会儿，尽兴而返。"

那必定是人家的花园。

"过两日，放了学又去，只见已有人在，我不顾三七二十一，拉着架子，就要站上，忽然之间，面孔上挨了一巴掌，金星乱冒，又被人痛骂一顿，只得知难而退。”

荷生动容。

“过数天，我再去。”

荷生惊愕，他自小是一个这样的人，永不放弃。

“这一次，我看到白衣黑裤的女佣在推一个小女孩坐秋千，那女佣很婉转地对我说：‘这是私家地方，不是你可以进来的，走吧。’”

荷生怔怔地听着，他不外想让她知道，他也受过羞辱。

“我终于走了，以后没有再去。”

荷生双眼润湿，她明白了他的一番好意。

烈战胜笑笑，“后来，我也建了好几座私人花园，却并没有设秋千架子，不过那热辣辣的一巴掌，至今难忘。”

荷生问：“打你的人是谁?”

烈战胜想一想，“是一个十四五岁穿唐装衫裤身形粗壮的女孩子。”他大概永远不会忘记她。

荷生点头说：“那家的打工妹。”

“我猜想也是。”

“当时你有多大?”

“七八岁。”

荷生气平了，笑出来。

"我一生受过不少挫折，皆能忘怀，大概无论什么事，第一次最难应付。"

"谢谢你。"

烈战胜面孔上打着问号。

"这个故事的寓意很好。"

司机把车停下来。

烈战胜送她下车，抬头看看天空，"明天会下雪。"

荷生茫然，她不懂天象。

烈战胜缓缓伸出手，轻轻抚摸荷生的面颊，随即放开。

荷生却如遇雷击，退后一步。那感觉，他的手指一碰到她的脸，她便顿感一阵酥麻，她认得这种震荡，她记得它不曾真正发生过，但却在梦中经历无数次。

她呆呆地看着烈战胜。

错了，不可能会是他，她实在太疲倦太焦虑了。

荷生匆匆掏出钥匙开门进屋。

关上门，脚下又是一封退回来的信，荷生弯下腰，疲倦地拾起它，丢在桌上。

她没有更衣，躺在床上一会儿，睡着了。

醒来的时候，枕头湿濡濡的，荷生将它翻到另一边，仍然赖在床上。

门铃在这个时候响起来。

荷生只得披上外衣去应门。

下雪了，一如烈战胜所预料。

门外是言诺。

荷生说："不要解释，一切都是我的错。"

这是维持人际关系最好也是惟一的办法：原来贤的是你，错的是我。

言诺站在门口说："荷生，你愿意嫁给我吗？"

荷生并不觉得意外，"进来再说。"

昨夜那件事完全激发了他的同情和怜悯之心，言诺放弃睡眠，与母亲吵了半晚，另外半夜用来伤怀。

言太太甚为震惊，她的孩子是好孩子，从来未曾使父母不快，一定是这个不祥的女孩子作祟，于是她更加进一步表明立场，"她要进门，我走。"

言诺马上说："不，她不会进来，因为我可以走。"

他真的走了出来，身边有件小小行李，装着简单的衣物。

他对荷生说："我没有地方可去，想在你处借宿。"

小公寓只有一间睡房，客厅没有沙发，只有一只睡袋，要是他想打地铺，或许有商量余地。

"喝了这杯咖啡，或许你改变主意。"

"我不会，第一次与你约会，我就已决定要你。"

“言诺，当中发生了许多事。”

“这些事也已经过去。”

很多人不会这么想，言诺的母亲是其中之一。

奇怪，人人都以为社会风气真正开放了，以前所计较的细节，今日都可以放过。

但不，一旦发生在自己身上，反应一样激烈。

荷生可以猜想假如言诺失去控制的话，言伯母随时会同爱子登报脱离关系。

荷生说：“要是你愿意，你可以在厨房露营。”

“没有问题，这已是我的最佳归宿。”

荷生看着他，“你会伤你母亲的心。”

言诺握住荷生的手，“在人生漫长的旅途中，总有些人有些心会伤害到你我或是被你我伤害。”

荷生正想笑着对这句话评论一番，忽然之间，掩着胸口，把适才喝下去的咖啡全部喷吐出来，言诺连忙抓起毛巾替她拭抹，荷生脸色苍白，伏在桌上喘息。

“你身体不妥，来，披上大衣，我同你去看医生。”

“不用麻烦你。”

“荷生，我同你之间，说这种话做什么?”

他挟持着她上车，找到医务所，上去挂号排队候诊。

医生给荷生做过简单的诊治，抬起头满心欢喜地对言诺说：“恭喜你们。”

言诺立刻明白了，他意外地看着荷生。

只听荷生镇定地说：“可否请大夫荐我去看妇产科。”

“当然。”医生写出单子。

两人道了谢，缓缓走出医务所。

言诺不知如何开口才好。

过了很久很久，他问：“你打算怎样告知他们?”

荷生失笑：“是我的孩子，何用告诉别人?”

言诺问：“你肯定你要他?”

荷生答：“已经考虑了整整四个月。”

言诺吁出一口气，“那么让我帮你。”

“我会害苦你的。”

“荷生，情形不会比你离开我那一天更惨，你放心。”

荷生摇摇头，“我愿意独自承担这件事。”

“我只不过在一旁协助，非必要时不出手，荷生，我并不打算把肚皮借出来。”

荷生觉得漫天风雪，大难当头，言诺还能拥有这一份天真，实在可贵，她笑出来。

言诺拉住她的手，百感交集，“我只希望有人爱我，如你爱他那么多。”

荷生微笑，“也许这不过是一个最最愚昧的选择。”

“我们回去再说。”

言诺为荷生预备简单的午餐，一边批评公寓不够大，最好有两个房间，不，三个房间，空气要流通，屋后要有玩耍的空地。

荷生坐在窗前，一言不发。

这时她听见门底“刷”的一声，转头一看，是邮差送信进来，她的信封，她的手迹——又一封退信。

荷生没有拾它，让它躺在地上。

言诺在厨房里自言自语：“搬了房子，就该准备一切了。我们要去找有关书籍来读增加常识，同时托人介绍个好医生，你要保持心情愉快。荷生，荷生?”他探头出来。

荷生坐着不动，窗外的雪越下越大。

言诺喃喃地说：“活像西伯利亚。”

他过去拾起退信，放在荷生面前，过一会儿问：“怎样把这个消息告诉烈火?”

荷生平静地问：“为什么要告诉烈火?”

言诺不敢再提。

“你千万别乱讲，我会不高兴的，暂时我不想让别人知道这件事。”

言诺温柔地说：“但是几个月之后每个人都会知道。”

荷生坚决地说：“以后再说。”

言诺问：“你不想增加他的心理负担，可是这样?”

“这件事与他没有关系，你别想歪了。”

“荷生——”

“我不想再进一步讨论这个问题。”

“当然，”言诺低下头，“我尊重你。”

荷生吁出一口气，“言诺，气消了就回家吧，伯母会挂念你的。”

言诺微笑，“我情愿留在此地，二十三岁的男子拥有自主权了。”

“有人会觉得你傻。”

他没有回答，打开了睡袋。

第二天，言诺出去办公，顺道送荷生到医务所。

荷生有点疑心她走起路来颇为蹒跚，但又不得不到城内商场买几件用品，返回公寓，觉得疲倦，靠在电梯口休息。

“夏小姐。”

荷生抬起头，是言伯母。

她穿得很庄重，黑嘉玛大衣，高跟鞋，可见是特地来探访她的。

荷生轻轻地说：“伯母，你以前是叫我名字的。”

言太太叹口气，“那时候怎么一样。”她也怀念那段日子。

“有什么不一样，我仍然是夏荷生。”

“荷生，你是不是要等烈火出来？”

“是。”

“那么，为什么要利用我的言诺?”言伯母开门见山，老实不客气地问。

“伯母要不要坐下喝杯茶慢慢讲?”

“言先生在车里等我。”

“言伯伯可要一起上来?”

荷生本与言家诸人极熟，此刻因无所求，问心无愧，更加坦然无惧。

言太太看着她，“我只有几句话要说。”

她跟荷生进家，在狭小的客厅坐下。

荷生斟杯热茶给她，为她脱下大衣，小心接好。

言太太开口，“荷生，我一直喜欢你。”

“是的，我知道。”

她痛心地说：“你太不自爱了。”

荷生忍不住，侧着头偷笑起来，如此陈词滥调，如今难得听到。

“我要你离开言诺，他有大好前途，快要订婚了，你不能自私耽搁他。”

荷生微笑道：“我何尝不是这样想，言伯母你这番话简直说到我心坎里去了，你劝他回家吧。”

言太太惊疑地看看荷生，“你不爱他?”

“我待他如兄弟，他是我的好友。”

“你不会缠住他?”言太太不放心。

“那种技巧，我一直没有学会过。”荷生向她保证。

“他现在何处?”

“上班去了。”

荷生这样合作，不外乎是帮助这位母亲减轻焦虑。

或许她十分过分，或许她侵犯他人隐私，或许荷生可以撵她出屋，但无论如何，她这一切所作所为，都是为了言诺。她是一个好母亲，正如所有好母亲一样，她认为孩子即使已经成年，但一旦失却她的厚爱保护，照样会化为一摊浓血。

言太太却认为夏荷生甘心听她教诲，乃是因为理亏的缘故。

她说：“当初你不该离开言诺。”

荷生没有说话。

“你有没有后悔过?”

荷生回答：“我没有时间后悔，不知道会不会后悔。”

言太太异常固执，“你会后悔的，放弃这样好的男孩子，你一定会后悔。”

这时，荷生觉得热，她站起来，脱下大衣。

言太太到底是个有经验的过来人，她注视荷生片刻，大惊失色，“你，你有了孩子!”

既然被她看出来，荷生点点头，“是。”

“谁的孩子?”她指着荷生。

荷生笑笑，“我的孩子。”

“这孩子是烈火的吧?”

荷生失去了忍耐力。

她取下言太太的貂皮大衣，“言伯伯在楼下等你好久了。”

言太太大惊失色，“我不准你再见言诺。”

荷生把大衣搭在她的肩膀上，拉开公寓大门。

“你别妄想把这宗烂账转嫁到言诺身上，我们祖宗积德，我们不会遭此污辱。”

她的爱至为狭窄，自家的孩子尊若菩萨，他家的子女贱若泥斑。

最不幸的是这也算人之常情，经历过大灾难的夏荷生已经不会为这等小事愤慨、激动、难过了。

她看着言太太离去，关上门。

后悔。

周末放了学，到言家去吃饭，同言伯伯下棋，吃伯母做的点心，每次他们都为她备下小礼物，他们欢欣地等她嫁进去成为一分子。

这样的结局当然幸福。

但后悔又是另外一件事。

荷生的脑海里没有这两个字。

中午时分，言诺返来，买了一大叠育婴指南书。

他又同房屋经纪人联络过，在稍远一个叫爱德华王子镇的小埠，有一幢平房，简直是建立新家庭最理想的地方。还有，他向校方打听过了，孕妇也可以照常上课。

言诺兴奋地说个不停，一洗过去沉实的本色。

“试想想，烈火一出来便会看见……”

“不要把烈火牵扯在内。”荷生再三警告他。

言诺只得搭讪地拿起一本画册，“《育婴秘方》，为什么不叫‘育英秘方’，培育英才嘛。”

“言诺，或许我们应该谈论一些比较现实的问题。”

“像什么？”

“言伯母今早来过。”

言诺放下书，“我已告诉她，请她不要多管闲事。”

“对她来说，这并不是闲事。”

“她讲了许多可怕的话吧？”

“没有我不能应付的话。”

“把摘要告诉我。”

“不用了，她肯定会对你重复一遍的。”

“你何必招呼她？”

“伯母也曾多次招呼过我。”

“不要记住这件事，下午去看新房子。”

“言诺，这是另外一个重要的问题，我身无长物。”

“我有。”

“不可以这样。”

“你不接受我的帮助，烈先生自会插手，我们断不能坐而不理。”

荷生微笑，“看样子，我始终是个幸运儿。”

“那我就不知道了，荷生，你也不必自嘲。”

荷生摇摇头，“我并没有不满意这间小房子，请勿安排我的生活。”

“你那犟脾气几时才改！”

荷生说：“我还有一个请求。”

“那又是什么？”

“回家去，这里住不下客人，你有空儿来看我即可。”

言诺静下来，过半晌他说：“看样子我的说服力还不如我母亲。”

荷生莞尔，“差远了。”

言诺吁出一口气，“晚上我来看你。”

“请你顺带替我寄出这封信。”

言诺接过荷生惯用的白信封。

一天一封，风雨无阻，再寄一千封，烈火也该出来了。

这封信，一定会落到烈火手中，纵然不拆开，单凭信封，也知道其中意思，内容已经不重要，也许收信就是烈火的寄托，也许他盼望不再收这样的信。

言诺找个轻松点的话题，“信里都写些什么，可以告诉我吗？”

荷生不语。

“你放心，他终究会拆开这些信的。”

荷生低下头。

“让我替你寄出去，莫使信链断开。”

他披上大衣走了。

言诺说的话总有他的道理。

荷生沉思良久。

没有人会知道，信中内容，有时抄自莎士比亚十四行诗选。

荷生有点饿，她去做了一份花生酱三文治吃。

这个时候，她真需要言诺这样的朋友。

天黑了，她没有开灯，心头如压着一块大石，花生酱全黏在嘴巴里，要用开水咽下去。

抽屉里一共有八十二封退信，尚有七封，迟早会抵达她的家门。

门铃响，荷生以为是言诺，待她洗干净双手，打开大门，看到的竟是烈战胜。

荷生站着不动，他一定是得到了消息，才来找她。

烈战胜留意她的一举一动，已经不是一段短时间了，像现在，他静静地站在门外，凝视着夏荷生。

荷生迟疑一下，挂上笑容，迎烈战胜进来。

客厅只有两张椅子，他挑了其中一张坐下，身材高大的他与小型家具格格不入，双腿简直没有地方放。

他喝一口荷生给他的咖啡，皱起眉头，他说："味道似焦米汤。"

荷生道歉。

他纳入正题，"琪园已经装修过，花园与停车场搬了位置，下个月烈云也许会搬回去住。"

"别叫她回琪园，太残忍了。"

"琪园届时不再叫琪园，会恢复叫落阳道一号。"他停一停，"荷生，你也回来吧。"

荷生摇摇头。

烈战胜温和地问："你为何强迫自己吃苦，你究竟想赎什么罪?"

荷生无言以对。

"荷生，首先我要替你搬一个地方，然后让你考虑清楚，什么时候返回烈宅。"

"你没有权摆布我。"

"我不是要摆布你，你的胎儿是烈家的人，我有权为他安排比较舒适的生活，相信你承认他是生命，相信你不会反对。"

"我的孩子与烈家无关。"

烈战胜沉默一会儿，"原来如此，"他说，"那么，你能不能接受一个长辈的一点心意?"

"我自己会处理。"

"如何处理?"他很直率地问。

"我会与家母商量。"

"她一直以为你已与言诺重修旧好，最新消息：

她已将你们祖屋变卖，资金当股份注入中华料理店，她不打算再回去了。”

“那更好，我可以名正言顺地去店里帮忙。”

“这个时候？”

荷生保持沉默。

“荷生，容我帮助你。”

“代价是什么？”

烈战胜微笑，“我并非慈善家，但很多时候，我都不讲条件。”

荷生小心翼翼地说：“烈先生，话还是讲明了的好。”

烈战胜不语，夏荷生开始有心机，他不可造次。

荷生问：“你想得到这个婴儿，是不是？”

烈战胜沉着应对，“依血统他是烈家的人，我何用费力争取他？”

“但，或许你想把他放进你所设计的人模子里去，自幼训练他成为你理想中的人物。”

烈战胜答：“很多人都是这样培养下一代的，你认为有什么不对？”

“我只想让下一代快乐。”

烈战胜抬起头来，“成功，或许，但快乐，未必。”

夏荷生战栗，他预言了胎儿的命运。

“荷生，上主最公平不过，生在我们家的孩子，拥有的固然不少，但失去的，也更多。”

“我要他做一个平凡的普通人。”

“在马槽出生的某人结果成为万世巨星，即使你是孕育他的母亲，你对他的命运也无能为力。”他停下来，笑一笑，“况且，你何尝不是意图把他套进你的模子里去，迫使他隐姓埋名？”

荷生认为烈战胜说得对，他们两人都过分偏激，可怜的婴儿，生活操纵在专横自私的成人手中。

天色已经全黑，荷生猛地想起来，“言诺呢，他在哪里？”

“我临时差他去见一个客人。”

他把言诺支开，好来与她谈判。

“相信你已猜到，他母亲来见过我。”

荷生莞尔，“声泪俱下？”

烈战胜点点头。

“她为一件不可能发生的事过分担忧。”荷生说。

“是吗，”烈战胜深意地说，“我们不应低估她的预感。”

他一直没有再碰那杯看上去像洗碗水似的咖啡。

他站起来，揉一揉发酸的膝盖，然后说：“准备明天搬家吧。”

荷生微笑，“可以看得出，烈先生，你急需一个接班人。”

烈战胜暗暗吃一惊，不动声色，也不再叮嘱什么，他走了。

言诺仍没出现，烈战胜差他到什么地方去了?

荷生翻开《育婴指南》第一章，字体渐渐模糊，她连忙揉揉眼睛，把忧伤的情绪压下去。

这个时候，她感觉到腹内一动，荷生愕然，她从来未曾有过这等奇突的感受，连忙站起来，吓得退至墙角。

接着腹腔内又似轻轻转动一下，荷生睁大眼睛，她忽然明白了，这是那小小胚胎，他开始在有限的空间内尝试着活动。荷生眼眶中泪水充盈，她缓缓低下头，双手轻轻覆在腹上，轻轻地说：“你好。”

他似听懂了，蠕动一下，作为回应。

荷生豆大的泪水终于重重滴下，她内心充满欢欣，几天来的疑虑一扫而空。试想想，她居然曾经考虑不要他!

荷生轻轻挪动身体，缓缓走到椅子前，坐好，在这一刻，她觉得自己珍贵无比。

电话铃响起，荷生才渐渐回到现实来。

“荷生，我是言诺。”

“你在哪里?”

他答：“烈先生有事令我到麻省走一趟。”

荷生一呆，无端端竟差他去那么远的地方。

“我刚看过烈云，情况令人宽慰，我明天中午可以回来，届时详谈。你可觉得寂寞？”

“不，我不觉孤独，”荷生说的是实话，“别忘了我们有两个人。”

“早点睡。”言诺笑了。

这个时候，荷生忽然发觉，她耳畔持续已久的“嗡嗡”声忽然之间完全消失，她可以清晰地听到钟的滴答声，她吁出一口气，这是不是从头开始的征象？

她斜躺在床上，闭上眼睛，心安理得地睡去。

第二天一早，夏太太来电找女儿。

“三刻钟的车程而已，荷生，你应该多来看我。”

“我正忙，也许要搬家。”

“荷生，能正式结婚，还是正式结婚的好。”

荷生觉得母亲的声音遥远，陌生，语气与腔调与她此刻的生活处境格格不入，宛如太空舱内航天人员与地面通话似的。

荷生不敢把真相告诉母亲，怕老式人受到刺激，但又不能想像在明年初夏某一日，突然拨一个电话给她说：“妈妈，你已荣升为外婆。”

荷生问母亲：“你的新生活如何？”

“过得去，忙得不得了，流汗流得非常畅快。”

“好！”

“有空儿同言诺一起来，记得了。”

荷生如释重负，抹一抹额角的汗。

她轻轻地说："那是你外婆，将来会疼你。"

她在厨房调制麦片，抬头朝窗外看去，发觉观点角度与前一日的她有太大的分别。

她开始有较长远的计划：孩子出生之后，她会带着他去投靠母亲，自力更生，把他养大。

最近身受一连串的苦难，都似被这一股欣喜淹没。

中午时分，烈战胜来接她，"荷生，新居已经完全准备好。"

"烈先生，我们在这里很开心。"

"至少来看看我的一番好意。"

荷生拉住他，"言诺几时回来？"

"他今天还有事要办。"

荷生看着烈战胜，即时明白，他是不想言诺在一旁影响他的决定。

"好的，我去看一看即返。"

那所平房宽敞舒适，设备齐全，其中两间睡房作纯白色设计，堆满各式婴儿玩具用品，有些箱盒尚未拆开。

荷生表示极其欣赏。

烈战胜问："你仍有犹疑？"

"我在想，中国人说的英雄莫论出身，不知是否有理。"

"这便是他的出身。"

"烈先生，你像是忘记了，他是我的孩子。"

烈战胜似有一丝恼怒，但一闪即过，他若无其事地吩咐："把钥匙交给夏小姐。"

"烈先生，我不能接受。"

他取过荷生的手袋，打开，把钥匙放进去。

荷生怕他下不了台，打算过两日再把钥匙还给他。

烈战胜建议，"我们一起午餐如何?"

"我已经约了人。"

荷生一而再，再而三地拒绝他，让烈战胜知道，他行动也许太过激进。

他只得退一步说："我送你回去。"

门口的台阶上有青苔，荷生走得小心翼翼，正在这个时候，有人伸出一只手来扶她。

荷生抬起头，好一个意外，"言诺，你回来了?"

烈战胜的意外程度并不下于荷生，他十分错愕，没想到言诺会擅自返回。

"烈先生，下午那个会，我已交给史蒂芬赵了。"

烈战胜问他："你到这里来干什么?"声音中怒意可闻。

言诺一怔，"我来接荷生。"

"我不记得叫你来过这里，荷生可以坐我的车子。"

荷生连忙说："是我叫言诺来的，我一早约了他。"

她设想到烈战胜会同言诺起冲突，急急挡在两个人当中。

她接着说："言诺，我们走吧。"

她拉着他的手上车。

车子驶离之后，荷生才笑笑说："恐怕要害你被老板责骂了。"

言诺看她一眼，"你开玩笑，刚才那件事已经足够令我丢掉工作了。"

荷生意外，"那么严重？"

言诺点点头。

"那你不该造次。"

"我不放心你。"

"言诺，"荷生实在无法不感动，"我可以照顾自己。"

"烈火不这么想。"

"对，他叫我与你结婚，你可知道比这更加荒谬的建议？"

"他知道他可以信任我。"

荷生忽然笑起来，"你们并不要我，你们要的只是我的孩子。"

言诺不出声。

"我关心烈云，请把她的事告诉我。"

"她认得我，还问我，荷生姐在何处？"

"言诺，我想跟她通话。"

"但是你要有心理准备，她只记得人，不记得事。"

是的，荷生黯然地笑，烈云连"快乐王子"的燕子都记得。

"还有，烈先生不是轻易接受拒绝的人。"

这是言诺第一次谈到烈战胜的真面目，荷生静待他说下去。

"烈先生有非常强烈的占有欲，成功本身便是不停占有，知足的人往往止于某处，极难出人头地，烈战胜对事业对家人都要百分之百控制。他从来不会鼓励烈火与烈云努力事业，可见一斑，他的爱与恨，同样炽热。"

他的子女也像极了他。

第二天，荷生出去找工作。

她看到图书馆里贴出招请临时工的广告已有一段时间。

荷生一出现，管理员如释重负，"这是一份闷坏人的工作，希望你会做得久。"以往他们每星期换人。

把破损的书页补起来，每小时的酬劳是十五元。

工作的环境倒是十分清静优美，不久，工作人员习惯了这个面貌清秀的东方少妇默默地坐在一个角落努力操作。

言诺来看过她，并且玩笑地说："别心急，做得太快，你会失业。"

其他人都以为他是荷生的爱侣。

荷生问："烈先生有没有责备你？"

言诺侧着头想一想，"没有。"好像有某种预感。

荷生答："那好，请你代我把这串钥匙还给他，我不打算搬到那所房子里去。"

言诺不语，他仿佛有点担心，"据我所知，他不会在此刻放弃。"

言诺说得很对，过两日，荷生抵达图书馆的时候，看到烈战胜坐在她的位子上。

"早。"荷生说。

"你是乘公共汽车来的？"

"不，我步行，可以省一点。"

烈战胜震惊地问："你以为我会任你过这种生活？"

荷生坐下来，握住烈战胜的手，"烈先生，你一直待我很好，一直给我自由，请不要在这个时候放弃对我的关怀。"

"你为什么不听我的安排？"

荷生正欲用最简单的言语解释她要独立的意愿，烈战胜却已经问她："是受了言诺的影响吧，他一直在等机会。"

"不，不是他，你千万不要误会。"

“我一直以为你与他早已没有纠葛了。”

荷生按住烈战胜，“听我说，这与言诺完全无关，我只想过自己的生活，你也已经默许了我。”

“现在情况不一样了，荷生，把婴儿给我，你可以走到天涯海角。”

荷生不置信地看着烈战胜，她终于见到他专横的一面。

“我不会把我的孩子交给任何人。”

“我不是任何人，我也不会任由你带着孩子嫁到言家。”

“我不想再说下去，烈先生，我要开始工作了。”

烈战胜在盛怒中站起来，一手把桌上一大叠硬皮书扫到地上，静寂的图书馆中发出震天的巨响。

他说：“我不会就此罢手！”

众人转过头来错愕地张望。

烈战胜已大步踏出，他身上大衣扬起，如一件张牙舞爪的巨氅，充满威胁感觉，他卷出大堂，用力关上门。

荷生受了震荡，跌坐在椅子上，突然感到胎动，她连忙说：“没有事，别怕。”

工作人员前来问候：“你不要紧吧？”

荷生摇摇头。

她慢慢蹲下，把书本逐一拾起。

荷生见过他炮制出来的人版，烈家三兄妹是最好的榜样，她怀疑在他屋檐下难以有真正快乐的人存在。

这种不快会互相传染，荷生后悔态度太过强硬。

那天晚上，言诺来看她，一进门，他便说："荷生，我有事与你商量。"

荷生奇道："我也正想说这句话。"

"那么你先讲好了。"

"不，言诺，你先请。"

"荷生，烈先生要派我做一件事。"

荷生心一跳，"那是一件什么样古怪的事?"

"他要我护送烈云返家。"

"可是烈云尚在接受治疗。"

"荷生，你想到些什么，不妨与我直说。"

"我想到许多许多，很远很远，我像是忽然开窍，以前所不明白的细节，此刻一一解开了。"

言诺脸色凝重。

"言诺，你试想一想，陈珊女士怎么会让烈云返回琪园，她好不容易才把女儿带出来。"

言诺抬起头，沉吟良久，"烈先生说，他付出了很大的代价。"

荷生说："这不算，我们每一个人都付出了代价。"

言诺托住头，"那又是为了什么?"

"有人要你暂时离开此地。"

言诺笑起来，“我不懂，谁会施调虎离山之计?”

荷生看着言诺，“你不妨猜一猜。”

言诺终于说：“荷生，那是一个非常大胆的假设。”

“是吗，言诺，在你心底下，你敢说你从来没有这样怀疑过?”

言诺不出声。

“他一直有个想法，我系受你唆使，正如他一直以为，烈云受着烈风摆布一样。”

言诺站起来，“荷生，你想得太多了。”

荷生说：“他并不容许前面有障碍存在，只是他没有想到，在清除异己的时候，连带牺牲了烈火与烈云。”

言诺说：“我相信烈先生出自好意。”

荷生叹一口气，他自幼尊敬烈战胜，他信任他，也是人之常情。

若今日图书馆那一幕没有发生，荷生不会这样烦恼，在烈战胜的逼迫下，她很自然要做出抗拒的反应。

“烈先生现在要我听令于他。”

言诺在狭小的客厅踱步，“他至少应该知道，夏荷生不是一个容易妥协的人。”

“真可惜，这个秘密只有我们两个人知道。”荷生笑。

“荷生，你需要休息，在图书馆里，先挑童话故事修补，此刻你不适宜看推理侦探小说。”

荷生问：“你会不会去？”

言诺考虑，“我会先了解一下真相。”

荷生把言诺送到门口。

言诺转头问：“医生有没有说是男孩还是女孩？”

荷生微笑，“重要吗？”

“女婴多可爱。”

荷生一直保持着那个笑容。

言诺一走，她便关熄灯火。

她知道有人在监视她的一举一动，因为她身份特殊，因为她知道得太多。

自从认识烈火那一天开始，她就有这个感觉，现在证实并非因她多心。

言诺第二天一早就拨电话到麻省陈女士公馆。

他要经过一番内心挣扎才能做出行动，以前，他从来没有怀疑过烈战胜。

他接到的指示是要在第二天下午抵达麻省陈府，自陈珊手中接走烈云，回家旅程已经安排好。

陈宅的电话很快接通。

言诺要求陈珊女士说话。

那边的管家却道：“陈女士出门了。”

“她什么时候离开的？”

“你是哪一位？”

“我是烈战胜先生的助手。”

“陈女士出门就是为了到纽约与烈先生会合，许是班机延误？”对方说，“你们可以查一查。”

言诺乘机问：“烈云小姐可好？”

“她情况一如以前。”

“谢谢你。”

言诺轻轻放下电话。

陈珊根本不在家，怎么会有人肯把烈云交给他。

荷生起码已经猜对了一半。

假如他依着指示准时抵达，必须留在当地，等烈云的母亲回来，那可能是一天，或许是两天之后的事。

烈战胜为何要调走他？

言诺深觉讶异。

当初，让他过来陪着夏荷生，也是烈战胜的主意。

彼时他已不在乎这个外人，他甚至不介意制造机会让她与旧男友重修旧好。

现在，事情完全不一样了。

烈战胜前后判若两人。

言诺看看时间，荷生在这时候应该抵达图书馆了。

他猜想得不错。

夏荷生在图书馆附近的咖啡室吃早餐，这两日她吃得比较多，肚子饱的时候有种非常满足的感觉。

今早，荷生觉得也该是把消息告诉她母亲的时候了，只是，该怎么开口呢，真是难。

这个时候，她最宽大的衣服也显得有点紧，面孔圆圆，增长的体重似乎有一半囤积在那里。

刚预备结账上班的时候，荷生一抬头，看到了她的母亲，她怀疑眼花，揉一揉双目。

可不正是夏太太。

夏太太静静地坐到女儿对面。

荷生十分讶异，“你是怎么找来的?”

“烈先生告诉我的时候，我还不相信。”

荷生看看表，“我工作的时候到了。”

“荷生，你应该让烈先生照顾你。”夏太太的声音相当镇静，“一个独身女子流落在外，有许多不便。”

荷生笑笑，喝一口咖啡。

“烈先生让我来劝你。”

荷生说：“看样子，娘家已经不欢迎我了。”

“你肯搬回家来吗?”

“我不想增加你的麻烦，”荷生说，“现在我这个身份，相信到任何地方去都不会太受欢迎，等几个月再说吧。”

“烈先生说你的倔强令他恼怒。”

荷生说：“他的专横也令我难堪。”

这个时候，言诺赶到了，他付了茶资，笑着说：“荷生，你去办公，我陪伯母谈谈。”

他永远是夏荷生的救星。

荷生如蒙大赦，披上外套，一溜烟儿逃走，动作依然灵敏。

夏太太感动地看着言诺，“你一直在照顾她吧?”

言诺说：“我们是好朋友。”

夏太太抱怨，“你不应放她走，生出多少事来。”

言诺安慰伯母，“即使如此，她也心甘情愿。”

“我真的担心她。”

“不用费神，荷生知道她在做什么，她比我们都理智勇敢，我对她充满信心。”

夏太太苦笑一下，“你总是帮着她。”

“相信我，伯母，荷生是一个非常特别的女子。”

这边荷生出了咖啡室，走到街角，看到一辆黑色大车停在前面，她看了看车牌号码，慢慢走过去，伸出手，敲敲后座车窗。

紧闭的黑色玻璃窗过了一会儿落下来。

后座位子上却不是烈战胜。

荷生仍然对那陌生人说：“请回去同烈先生说，他的一番好意我心领了。”

她缓缓走开。

中午，言诺来接，荷生笑道："难怪要支开你，你永远在一旁碍事。"

言诺陪她散步到公园，"伯母已经回去了。"

"我还得多谢烈先生，他免去了我对母亲的坦白之苦。"

"在这几个月里边，他仍然会不断努力。"

"也好，"荷生说，"这样他可以消磨时间。"

"你那要命的幽默感好像恢复了有五成以上。"

言诺说得对，荷生似已找到了新的力量。

荷生停步，"言诺，你去送烈云吧，这里我自己可以应付。"

"也该有人把这个好消息告诉烈火了。"

荷生看着足尖，"言诺，请勿违反我的意愿。对他来说，这不一定是好消息。"

"我明白你的顾虑。"

荷生说："言诺，试想一想，假使我们这些人统统没有出生过，上一代的生活岂非轻松得多？"

言诺不出声，过一会儿他问："孩子取什么名字？"

"不是有三天假期吗，趁那个时候，好好地想一想。"

言诺灵机一动，"荷生，我们可以一起去探访烈云。"

荷生心动，嘴里只说："被言伯母知道我俩一起行动恐怕又会触发一场误会。"

言诺瞪她一眼，心里却十分欢喜，荷生已大有进步。

第二天早上，荷生照常到附近的邮筒去寄信，猛地想起，昨日并没有收到退信。

她不希望这是邮误，她希望烈火已经把信收下拆开了。

她有点激动，连忙回忆那封信的内容，热泪盈眶。

中午，言诺来接她出门，她的心境犹未平复。

荷生提着简单的行李刚走到门口，已经有人过来拦截，荷生认得那人，她昨日在街角车厢内见过他。

那人一时情急，竟冒昧地问："夏小姐，请问你到什么地方去?"

荷生大大诧异，反问："你是谁？我为何要向你汇报?"

那人退后一步，连忙返回车内。

言诺与荷生出发去飞机场。

言诺看看倒后镜，"他跟在后面。"

荷生无言，可以想像当年也有人这样盯住烈云，这是何等的压力。

荷生忽然说："把车停到前面加油站去。"

言诺问："什么?"一边已经把车慢下来。

荷生吁出一口气，待车停下，她说："我去买一罐苏打。"

后面的黑色大车也跟着停下来。

荷生走过去，司机佯装看不见她。

荷生轻轻地说："我不打算开始逃亡的生涯，请告诉烈先生，我此行是偕言诺到波士顿探访烈云，我很安全，婴儿也安全，请烈先生莫紧张。"

司机听了荷生如此坦诚的一番说话，十分惊讶，脸容松弛下来，终于说："夏小姐，谢谢你，你使我的工作容易许多。"

"不用客气。"

荷生回到车子里，言诺问："你同他说了什么？"

荷生回答："原来走出迷宫的方法再简单不过。"

"说来听听。"

"只要伸手推倒面前的障碍就行，我们一直犯了大错，兜完一圈子又一个圈子，愚不可及。"

言诺开动车子，直到抵达飞机场才觉悟过来，他说："看情形你终于接受了烈先生。"

"是的。"

"并不容易。"

"我知道，但我想再斗下去也没有意思，我愿意做出适当的让步，希望他也会体谅我。"

言诺微笑，"烈先生只晓得进，不懂得退，商量一词对他来说，是由他告诉你下一步该怎么做。"

荷生说："发生了这么多事情，难道他仍然一成不变?”

“我不知道，或者你是对的，值得一试。”

在候机室荷生轻轻推一推言诺，言诺朝她暗示的方向看过去，只见烈战胜远远地站着，朝他们点点头，随即转身离去。

可怜的人。

将他的一生得失归纳一下，他过得极其贫乏。他的原配对他不忠实，他的长子并非由他所出，他与后妻感情破裂，烈战胜是悲剧中的主角。

奇怪的是，从来没有人从这个角度看过他。

言诺见荷生怔怔的，便在她耳边说：“他已经走了。”

荷生抬起头问：“他到什么地方去，他可认得回家的路?”

言诺一愕，“他是烈战胜。”

荷生随即笑了，“的确是，他是烈战胜。”

到达陈府，管家不让他们进去，守卫如此森严，可见是怕有人带走烈云。

言诺留下姓名及酒店电话后偕荷生离去。

荷生在一间人工湖畔的小餐馆内写明信片。

言诺以为她要寄给烈火，看到地址，原来是问候母亲。

荷生说：“我们极少照父母的意愿长大，三岁一过已经自由发展，各有各的命运，各有各的道路，难免叫大人失望。”

“夏荷生将为人母，感慨突增。”

荷生忽然想起来，“那位与你相亲的漂亮小姐呢？”

“她肯定我与旧情人藕断丝连，已经避不见面了。”

“为这样好的男孩子，她应当出来同我决一死战。”

“荷生，你总是高估我。”

荷生笑了，她拍打着言诺的肩膀，心中也承认，能把从前狭义的感情升华到今日这个地步，真不是一件简单的事。

傍晚，电话接通，陈珊女士愿意见他们。

她站在门口欢迎荷生，“我知道夏小姐一直是小云的朋友。”

荷生十分惭愧。

“请进来。”

大家坐好，寒暄过后，不知道如何开口，三人一直是面面相觑。

隔了许久许久，大家静静坐着，但空气中不知有些什么，使荷生的鼻子有点酸意。

终于，陈女士问：“最近有没有人见过烈火？”

他们摇摇头。

陈女士难堪地说：“他不肯见任何人。”她深深叹息。

会客室里又静下来。

还是陈女士打破沉默，“夏小姐，我去带烈云出来。”

烈云胖了，整个人看上去圆圆的，一见荷生，就把她认出来，趋到她身边叫：“荷生。”

荷生紧紧拥抱她，“烈云，你太好了，看，这是谁？”

烈云只是笑，“原来是言哥哥，请过来这边坐。”

她母亲脸上却没有欢容。

荷生过去说：“烈太太——”

“我早已恢复本姓。”她停一停，“结婚二十多年，真正做烈太太的时间，大约不超过一个月。我对丈夫并不认识，对子女甚为陌生，失败得不能再失败。”

荷生笑了，见到陈女士仍然率直如故，觉得快慰。

她接着问：“言诺，你可不可以告诉我你的老板搞什么鬼，约好我在纽约见面，却叫我扑空。”

言诺赔笑：“他另外有要紧的事走不开。”

“你可以同他老实地说，十六年前我把烈云交给他是我最大的错误，今天我不会重犯。”

荷生跟随烈云走到温室，烈云一转身，看到荷生，非常惊讶，“荷生，你怎么在这里？”

荷生陪她坐在长凳上，“我来看你。”心中明白，烈云已经失却记忆，任何事，转瞬即忘。

荷生知道她不该这么想，但又禁不住这么想，能够全盘忘却，是多么好的一件事。

正在感慨，忽有一股奇异的清香钻进荷生的鼻孔，她转过头去寻找香气来源，看到花架子旁放着一式两盆曼陀罗花，十个八个蓓蕾正盛放着，这香气勾起了荷生全身的七情六欲，她一生的悲欢离合纷纷繁繁，笑泪忽然都在刹那间涌上心头。

荷生忍不住，匆匆用手掩上面孔。

“荷生，”烈云问，“你怎么了？”

荷生轻轻答：“没什么。”

“荷生，你为什么哭？”小云握住她的手。

荷生答：“我思念烈火。”

烈云笑一笑：“呵，烈火。”

这时言诺唤她们，“小云要加件外套吗？”

荷生对烈云说：“我们回去吧。”

看护过来把烈云领走。

言诺过来，只看见荷生嘴角挂着一个暧昧的笑容。

他安慰她：“有朝一日，烈云会把前尘往事一一归纳起来的。”

荷生抬起头，“到时恐怕她会惊叫一声，痛哭失声了。”

言诺蹲下来，“这是什么话，我以为你已经振作起来了。”

茶点已经准备好。

陈女士说："荷生，我知道你一直想重组这个支离破碎的家庭。"

荷生讶异地说："不，我从来没想过要做力有未逮的事情。"

陈女士微笑，"你很快就会有得力助手。"视线落在荷生腹部。

荷生有点尴尬。

"真没想到今天有这样一件令人鼓舞的好消息。"

荷生问："你支持我？"

陈珊毫不犹疑地拥抱荷生，"我多愚鲁，要待言诺告诉我，我才注意到。"

"你做祖母是太年轻了。"荷生微笑。

"言诺说你打算自己照顾他。"

荷生点点头。

这时候烈云走近，"你们在说什么，好像很高兴。"

荷生伸手招呼她，"过来，蹲下。"

小云照荷生的指示把耳朵贴向她腹部，胎儿碰巧踢动一下，小云吓一跳，"哟，"她说："有人。"

言诺先大笑起来，"小云说得好，可不真是有人。"

烈云也笑了，她仍把双臂搭在荷生肩上。

那天晚上，荷生把这个笑话写出来，寄给烈火。

言诺问荷生："节目还称心吗？旅程还愉快吗？"

荷生答："我担心回去要看烈先生严厉的脸色。"

"你是我们当中惟一从来不理会他脸上颜色的人。"

荷生叹口气，"我不应那么做，我该对他好一点。"

第二天他们带烈云到公园喂鸽子。

看护与司机紧随着，荷生有点不自在，烈云却非常满足。

她如三岁奶娃似的满草地追逐飞鸟。

荷生忽然觉得烈家的孩子命运奇异，见得到母亲便见不到父亲，双亲犹如参商二星，不许团聚。

她轻轻地对胎儿说："你恐怕也要过一段这样的日子。"

言诺一直不离烈云左右。

吃完冰淇淋，他们送小云回家。

烈云在门口拉住荷生，不舍得她走，神情茫然，却想不出什么理由留住荷生，荷生恻然。

陈女士亲自出来道谢，"有空儿再来，保重身体。"

归途中，荷生对言诺说："你可以放心了吧，我已找到了新的力气。"

言诺点点头，"我很佩服你，荷生。"

"作为烈火与夏荷生的朋友，没有人能比你做得更好更完美了。"

言诺说："开头，我不是没有私心的。"

"向烈先生辞工吧，也许你应该回家陪父母亲，不然与长辈的误会日深，终有一天筑起一道冰墙。"

"现在轮到你安排我的生活了？"

荷生笑笑。

"有人说，最怕人家对他好，因无以为报。"

荷生默然，言诺已经为她牺牲得太多太久，的确是一种压力，他应该去开始新的生活。

言诺问荷生："你要我走？"

荷生点点头。

"好的，我走，不过别说我不告诉你，一回到家，我马上会开始大宴群芳的。"

荷生由衷地说："太好了。"

言诺沉默下来，"荷生，我想问你一个问题已经很久了。"

"我知道。"

"你晓得问题是什么？"

"当然。"

言诺不忿，"说给我听。"

"大学一年级欠下的英国文学笔记，到底打不打算还我？"

言诺看着夏荷生，一直笑，笑得眼泪掉下来，然后他轻轻吻她的额角脸颊，"夏荷生夏荷生，你永远令我倾倒。”

荷生不敢让他听见她的叹息声。

她当然知道言诺要问什么，他要问：荷生，从头到尾，你有没有爱过我？

她一直怕他终于忍不住会问出口，她不想说谎，但是内心深处，到现在，她明白了，自己不算真正地爱过言诺，因为假如有的话，她不知道该怎么形容她对烈火的感情。

车子停在门口，言诺对荷生说：“需要我的话找我。”

夏荷生回家推开门第一件事便是留意有无退信。

没有。

地板上光光滑滑，什么都没有，连电费单广告函件零碎单张都没有。

荷生松弛下来，沐浴更衣休息。

然后她发觉她还有一个舒服的原因，她走到客厅，拉开窗帘，直看到街上去，那种被偷窥的感觉到今日才算完全消失，监视她的人，已经离去，荷生希望他们以后都不要再来。

是夜荷生睡得非常好。

第二天一早她出门去上班，那辆再熟悉不过的黑色大车立刻驶到她面前，司机下车招呼她，“夏小姐你回来了。”

荷生点点头。

“今天要用车吗？夏小姐。”

“不用，我步行，反正需要温和的运动。”

出乎荷生意料之外，那司机递张卡片给荷生，“夏小姐，需要我的话，拨电话给我。”

他随即上车驶走。

多么文明！

荷生不相信烈战胜会给她这么多的自由，尊重她的意愿。

别看这小小的一项改变，对烈战胜这样的人来说，简直是艰难的一大步。

一整天都不再见有人前来谈判干扰。

伏在案上工作久了，颈项背脊都有点酸软。

中午去饭堂吃一客三文治，回来再做，一直到下班时分，都无人骚扰，荷生抬起头来，恍若隔世。

她喃喃自语，“孩子，再没有人来理我们了，不管我俩在这里自生自灭。”感觉非常矛盾。

荷生害怕她会一辈子坐在这个位子上为图书馆修补破书一直到白发苍苍。

原来一切在争取到自由后才刚刚开始，难怪有许许多多女性根本不去向往海阔天空，她们情愿伏在熟悉的巢穴中天天抱怨。

图书馆八时整关门，同事见她迟走便问她：“你身体没有不适吧？”

“没有。”

她收拾好杂物回家。

天色已经漆黑，荷生有退回室内拨个电话给司机的冲动，终于忍下来，自手袋取出一块巧克力，咬一口，努力向前走。

荷生听到有脚步追上来，连忙转身。

是适才那位热心的女同事，荷生又失望了，她满以为是言诺来接她。

“我们一起走吧。”女同事笑着说。

荷生点点头。

开头的时候，他们，包括她母亲，把她缠得奄奄一息，几次三番，荷生几乎在窒息的情况下想失声痛哭，现在，他们终于听从她的哀告，荷生又觉孑然一人的孤苦和可怕。

她仰头看到天空里去，只见到疏落的星，她内心有点悲凉，世上难道真无中间路线，抑或还待苦苦追寻？

女同事说：“我们一直嚷要独立，现在丈夫们乐得轻松，都不再来接送我们了。”

荷生只得笑笑。

女同事想起来，“我们好像见过你丈夫几次。”

荷生简单地答：“最近他比较忙。”

她俩走到一个路口，女同事说：“我要在这里转左，你好好当心。”

“对了，”荷生问，“这个冬季什么时候过去？”

“快了，树梢已经发芽，”同事笑，“第一个冬天的确难挨，不过我们的春季会使你惊艳的。”

荷生笑，“明天见。”

她慢慢走回家，一路上想到许多形容词，像踯躅，像蹒跚，像颠簸、像流离……

街角的面包店刚要关门，荷生还来得及进店去买最后一只葡萄干卷，店东同她熟，“还以为你不来了呢。”

荷生道谢。

“好好照顾那婴儿。”

她打开门，仍然没有退信。

她假设烈火已经把信件收下阅读，下一步，或许他会回她只言片语的。

目前荷生要做的是熟悉这种清淡的生活。

睡到半夜，她听到有人叫她：“荷生、荷生。”

又是那熟悉的梦。

她游离着自床上飘浮起来，追溯声音来源。

她看到有人背着她坐在客厅那张小小的椅子上，那人缓缓转过头来，她发觉他是烈火。

他面容沧桑许多，胡须头发已经清理过，他笑问荷生：“你还在等？”

荷生答：“是，我一直在等。”

她走近烈火，伸手过去，触及他的脸庞，感觉太真实了，荷生问：“你吃了许多苦吧？”

烈火点点头。

荷生心底下明知道这是一个梦，却也觉得十分欢愉，刚要进一步问候烈火，电话铃骤然响起来。

荷生的精魂遭此一惊，马上归回床上的躯体，她跃起来，掀起被褥，出去听电话。

太煞风景了，是谁有什么要事，急急要与她说话？

她看一看钟，才七点整。

那头是个外国人，荷生一听，啼笑皆非，分明是打错了电话，刚欲开口，那人却问："你还在等？"

荷生一怔，泪珠上涌，纷纷落下。

对方声音异常稚嫩，分明是个少年人，也只有十八九岁的大孩子，才会在晨曦拨电话问出如此傻气痴情的问题来。

荷生忍不住答："是，我一直在等。"

那边听到不是他期望的声音，只当荷生开玩笑，"咔"的一声挂上电话，听筒内只剩下呜呜声。

春寒料峭，荷生搭上一块披肩，坐在窗前，掩着面孔。

有人以为生老病死贫最苦，虽是事实，但思念之苦，也足以使人万劫不复。

静坐良久，她抬起头来，看到门外的樱桃树枝上果然已经附着点点绿芽。

十天之后，当这些嫩芽都生长伸展成为半透明翡翠叶的时候，荷生才再一次听到言诺的声音。

"身体好不好，生活如何？"

荷生十二分惊喜，“好家伙，我以为你要避我一辈子。”

他只是笑，“真正物以稀为贵，以前看到我一直有厌恶感，今日口气却如获至宝。”

荷生说：“一日不见，如隔三秋。”

“可是，荷生，你说得对，我们过去企图经营你的生活也太过分了。”

“言诺，现在连烈先生都放弃我了，司机保镖统统不再包围我。不是没有一点点遗憾的。”

言诺意外，“真的？没想到烈先生会这么做。”

“我们今天晚上能否一聚详谈？”

“呃——”

“言诺，不是每晚皆佳人有约吧？”

他笑，“荷生，我在家里，这是长途电话，只怕今夜赶不到你处赴约了，后天晚上如何？”

“你回了家！”

“是，父母与我已经冰释误会。”

“我真替你高兴。”

“芥蒂仍存，真没想到家母会这样蛮横。”

“嘘，当心她听见。”

“幸亏你不用嫁到我们家来。”

这时候，荷生隔着一个大西洋，忽然听到言诺那边有人莺声呖呖地问：“谁呀，谁不嫁给你？”

言诺有点尴尬，“荷生，那是——”

荷生连忙接上去，“你的英文补习老师？”

“不——”

“你的表妹之一，那是一定的。言诺，我们后天晚上一起吃饭吧。”

言诺一直赔笑，“要不要我带什么来？”

“要，烈火的消息。”

言诺沉默一会儿，“我尽力而为。”

大学人事部约见荷生，向她透露一个喜讯。

他们想聘她为永久雇员，那样，她可以享用医疗服务、产假以及其他福利。

荷生马上答应下来。

一定有哪个善心人替她递了推荐书，帮了她一个大忙。

是谁呢？

回到位置上刚坐下，那位女同事便朝荷生笑笑。

荷生明白了，她过去说：“谢谢你。”

“申请文书才递上去，还要看你履历经验适不适合，况且，这亦不是一份华丽的工作。”

“我衷心感激。”没想到在这里也可以结识到朋友。

“看得出，你本来不止过目前这样的生活。”

“不不不，我比较喜欢现在。”

“其中一定有个感人的故事，在适当时候你或许愿意告诉我。”

荷生微笑，重新回到位子上去工作。

如果想在这里落地生根的话，机会已经来临，可以把握。

她母亲是此地的永久居民，可以申请女儿入籍，并在此工作。

噫，多久没有处理民生问题了。

荷生这才发觉，无论如何，人原来都得活下去。

言诺带了一只小巧美味的巧克力蛋糕来看她。

荷生决定先吃一块再出去吃饭，谁知一块不足，又添一角，然后以为言诺没留意，再偷偷塞一口进嘴巴，足足吃了小半个蛋糕。

言诺没想到短短两星期内荷生会胖这么多。

她像是很满足很平和，这真令言诺伤心，他情愿她敏感而悲伤，他心目中美丽的女人，应该永远抱怨现实，处处感到不足，但是荷生仿佛已经习惯生活中种种不如意的挫折，甚至身为悲剧主角亦已麻木。

言诺一心一胸都是泪意。

刚在伤感，荷生却问他："你的表妹好吗？"

当晚电话旁的确是他的远房表妹，他不想解释，只答："好，谢谢。"

荷生又问："见过烈火没有？"

"烈先生正与律师商议明年保释的事宜。"

荷生已经猜到烈火仍然不肯见朋友，她低下头。

果然，言诺说："我只跟他说过几句话。"

"有无提到我？"

"有。"

"有没有好消息？"

言诺答："听他的声音，心境像是十分平静。"

荷生要求低，听了这句话，已经满足地吁出一口气。

"我们出去用晚餐。"

荷生问："言诺，时间是否真的能治愈一切忧伤？"

言诺答："可能会，但是如果要等二十年伤口才愈合，又有什么益处？"

言诺越来越成熟，越来越温和，与他相处，那种感觉就像喝下极其香醇的陈年佳酿。

荷生不由得说："你表妹是位幸运的女郎。"

言诺在荷生寓所楼下四处张望，果然没有见到烈氏派来的人马。

但是他了解烈战胜远比荷生深，他知道烈氏不会全盘放弃的。

他们一定还在附近，悄悄地执行任务，只不过略把行动收敛了。

言诺想起烈火同他说："我真不愿再给荷生任何虚假的希望。"

烈火的声音镇定而苍老，异常冷淡，提到夏荷生，像是在说陈年往事。

"荷生也需要精神支持。"

"我知道。"

"你应该回她的信。"

烈火没有回答。

言诺得不到答复，心里一酸，荷生那卑微的盼望又落空了。

烈火说："世上的确有从头开始这件事，最好她由她开始，我由我开始。"

"烈火——"

"谈话时间已经到了，再见。"烈火像是毫无留恋地挂上电话。

言诺这才发觉，烈火是多么的像他的父亲烈战胜。

荷生看到言诺对着丰盛的食物不能下咽，诧异地打趣："表妹同你有龃龉？"

言诺强笑，"她哪里敢逆我意。"

荷生觉得言诺越来越可爱，忍不住拍拍他的肩膀。

"现在，你可以告诉我了，烈火肯不肯见我。"

言诺轻轻地说："他仍然躲在茧里，不愿意出来。"

荷生忽然生气了，"他们两兄妹不约而同采取这种自私的方式来保护自己，却造成他人更大的痛苦。"

言诺只得三分同意，烈火的心情可以了解，他不想荷生继续为他牺牲。

他空肚子喝着酒，渐渐有点醉意。

荷生说："我们回去吧。"

"荷生，看样子你要独自熬过这个难关了。"

"我早有心理准备。"

话是这样说，荷生还是觉得气馁了。

隔日荷生怅惘地去医务所。

医生笑着同她说："是女孩子。"

荷生一怔。

"不喜欢女孩子?"

女孩往往比男孩更令父母担心。

医生说："我喜欢女孩。"

回到图书馆，女同事前来慰问："检验结果如何?"

"一切正常，谢谢。"

"那我要与你去庆祝一下，你还没有约人午餐吧?"

荷生微笑，"一言为定。"

谁知道她忽然说漏了嘴，"我也喜欢女孩子。"

荷生灵光一闪，电光火石间一切都明白了，她不禁哑然失笑，哪里来的那么多好心人，原来医生和同事都是烈战胜的手下。

但是这一次荷生却没有反感，她佯装听不出破绽，若无其事地做她的日常工作。

烈战胜比从前含蓄多了。

夏荷生也是。

女同事忐忑不安，试探荷生数次，荷生一点痕迹都不露出来，她们仍是朋友。

烈战胜煞费苦心，才做出这样的安排，荷生实在不忍心拆穿。

他们之间，已经产生了解。

荷生在下班时分，拨电话给他。

烈战胜决没想到夏荷生会主动与他接触，本来正与私人助理商讨一些重要事宜，也立即宣布休会，他问荷生："可是有要紧事？"

"没有，能不能一起喝杯茶？"

那口气，完全就像女儿对父亲一样自然平和。

烈战胜却受了极大的震荡，因为从来没有人这样对他说过话。

他清清喉咙，"明日下午四点，我来看你。"

"烈先生，明天见。"

荷生准备了茶点，又特地把一只书架子移到房中，使客厅宽敞一点。

她备下蒸漏咖啡壶，试喝过制成品，颇觉可口，才决定拿它来招呼客人。

听到敲门声的时候，荷生记得她看了看表，才三点三刻，她抹干手，去开门。

门外站着一个老妇，骤然间荷生没有把她认出来，她佝偻背脊，双手紧紧扣在胸前，最离奇的是她一把花发，分成两截颜色，前白后黑，原来染惯了头发停下来便会如此怪诞。

荷生并不认识她。

她也不认得荷生，因此她问："夏荷生在吗？"

"我就是夏荷生。"

"你就是夏荷生？"

荷生暗笑，这些日子来胖了十多公斤，但是，这是谁，她们以前难道见过面？

"你不记得我？"老妇抬起头怨怼地问。

荷生摇摇头。

"都过去了是不是，连琪园都忘记了？"

荷生一震，浑身汗毛竖起来，不可能，这不会是周女士，这名老妇看上去足足有七十多岁，怎么会是她？

荷生退后一步。

她抚摸着面孔，"我真的变得那么厉害？"

荷生慌忙答："大家都跟以前不同了。"

"是的，"她喃喃地说，"你也完全不一样了。"

"请进来。"

"你让我进来？"

"你不是来看我的吗？"

她点点头，"不错，烈风一直说，只有你没有偏见。"

荷生恻然，不忍看她。

"我来问你一个问题。"

荷生不顾三七二十一，抢了机会说："我也想问你一个问题。”

老妇凝视荷生，双目绿幽幽的十分可怕，“好，你先问。”

“烈风不是烈家的孩子，是不是？”

她被荷生占了先，十分不悦，但不得不拿她所知，来换她想知道的，她点点头。

荷生松一口气，她终于释疑了。

“轮到我发问了。”

“请问。”

“那件事，真是一宗意外？”

荷生点点头，“的确是意外，坠楼的可以是他们两人中任何一人。”

“你发誓？”

“我发誓。”

“以你腹中的孩子发誓。”

还是不肯放过任何人，但是荷生心平气和，她说：“我以我的孩子发誓，那是一件意外。”

老妇仰起头吁出一日长长的怨气，荷生听在耳中，只觉无限阴森，浑身皮肤起了鸡皮疙瘩，胎儿忽然躁动起来，不住地踢她。

荷生轻声安慰，“没有事，不要怕。”

但忍不住又退后一步。

“这么说来，你在法庭上没有说谎？”

荷生瞪着她。

“我走了。”

她站起来，颤巍巍地走到门口，打开门，离去。

荷生一直僵在角落里，过半晌，门铃再度响起，她方回过神来，看看时间，刚刚四点整。

她去开门，烈战胜吃惊地说："荷生，你脸色好坏。"

荷生连忙说："我一定是等急了。"

"荷生，让我再看看你。"

荷生忍不住，"烈先生。"

她把脸埋到他胸前，假如她有父亲，她也会这样做。

"你浑身颤抖，告诉我，是怎么一回事？"

烈战胜扶她坐下来，渐渐的荷生灰败的脸色才恢复一点点红润。

她忍不住告诉烈战胜，"我看到她了。"

"谁？"

"琪园的旧主。"

烈战胜吁出一口气，"那是你的噩梦，那人卧病在床，况且，即使你看见她，也不会认识她，她已经衰老不堪了。"

荷生更加肯定她没有看错人，"是她，我真看见她了。"

烈战胜的语气十分肯定，"健康情形早不允许她远渡重洋，那不可能是她。"

荷生知道他一时不会相信，只得斟出咖啡招

待。

烈战胜尝一口，“比上次那杯好多了。”

荷生笑一笑。

“你可是有话同我说？”

荷生低着头看着杯子，“一家人，也别太生疏了，烈火把我们拒绝在门外，我们又忙着制造纠纷，这样下去好像没有什么帮助，将来烈火看到这个情形，恐怕会失望。”

烈战胜讶异，“我不明白你的意思。”

荷生伸出手来，“让我们做朋友。”

这个女孩子的倔强，令烈战胜深感诧异，她毫不妥协，亦不愿听他摆布，但她愿意与他平起平坐，握手言和。

烈战胜只得伸出手来，他很清楚，只有这个办法可行。

“我知道身边仍然都是你的人。”荷生微笑说。

烈战胜有点尴尬，随即说：“我觉得你需要照顾。”

“我这才知道十五元一小时的工作也得靠人事成就。”

正渐渐谈得融洽，忽然有人敲门。

烈战胜问：“荷生，你在等人？”

荷生讶异，“不，我没有约其他人。”

她去开门，门外是她见惯见熟的那位司机，当然，到这个时候，荷生也很明白这位先生的地位断不止司机那么简单，他是烈战胜的亲信之一。

“夏小姐，请问烈先生在吗?”

烈战胜已经迎出来，“什么事?”

“烈先生。”他趋向前，在烈战胜耳畔说了几句话。

夏荷生看着烈战胜的面色骤变，知道这宗消息非同小可。

只听得烈战胜问：“什么时候的事?”

亲信又轻轻说了一句话。

要过半晌烈战胜才能说：“你先回去。”

然后他转过头来凝视荷生，荷生此时已经不再恐惧，她完全知道发生了什么事，她温和地问：“周琪女士方才过世，是不是?”

烈战胜点点头。

荷生心中明白，她只有一件事放不下，想知道答案，荷生已经把实情告诉她，她可以瞑目了。

“荷生，你说你方才见过谁?”

荷生镇定地说：“日有所思的缘故，我做梦了，刚才等你等得有点累，一定是睡着了。”

烈战胜知道她不肯多说，于是低头道：“我要替她去办理后事。”

荷生为之恻然，“我猜想她已经没有亲人了。”

烈战胜摇摇头，证实了这一点。

荷生问：“是什么疾病使她外形猝然衰老?”

烈战胜佯装没有听出破绽来，“癌症。”

荷生一直送他到停车场。

烈战胜问："荷生，你决定等?"

荷生答："不，我决定生活下去。"

惟有采取这样的态度，才能挨过这段日子。荷生并没有准备闲下来，她并没有打算看日出日落便当做一天，日日呻吟，夜夜流泪，她真的想正常生活。

"请告诉烈火，我并没有为什么人牺牲。"

烈战胜说："听说是个女孩?"

荷生微笑，"不论男女，你都会失望，我带孩子的方法，与烈家大有出入。"

"她会姓烈吧?"烈战胜还存有最后一线希望。

荷生非常坦白，"我不认为会。"

烈战胜十分气馁，"我希望你回心转意。"

荷生笑，替他关上车门。

"荷生，"他按下车窗，"我们有空儿再喝茶。"

"当然。"

他去了。

荷生回家，看到自己的影子，怀疑不速之客又来探访，蓦然回首，走廊却空无一人。

恐惧亦会用罄，一如眼泪，去到尽头，黑暗化作黎明，往往有出人意料的发现。

荷生时常怀疑烈风就在街角等她，她相信他会挑选树阴最最浓密之处，但枝叶再茂也遮不住他削薄的脸容，憔悴的大眼，瘦长的身段。

荷生相信在百步之遥便可以把他认出来。

好几次在黄昏穿过公园，她都仿佛看到他。

她趋向前去，轻轻地问："烈风，你在那里吗？"

她希望他会慢慢走出来，就像以前那样，似笑非笑地看着她，对她似有好感，但明明又是对立的一个人。

荷生比什么时候都想念他，假如现在才开始认识他，荷生会把关系处理得比较好，也许悲剧不会发生。

现在她只希望与他说几句话。

每日上下班她都故意走同一条路，等他前来相会，但她始终没有再见到他，或许他不再信任她，或许他对她不满，荷生觉得深深失望。

她的行动渐渐不便，母亲来探访她，仍然问："言诺呢？"夏太太永远不会忘记这个可爱的男生。

他两地穿梭，忙着事业跟学业。

夏太太说："他也不大来看你了。"十分遗憾。

"相信他已经开始新生活了。"

自母亲眼中，荷生猜到她想些什么。

母亲一定在想，烈火同言诺两个人，夏荷生明明认识言诺在先。

不知怎地，荷生没有嫁给言诺，但也没有嫁给烈火。

她落得孑然一人。

言诺终于抽空儿来看她的时候，并没有带来好消息。

“荷生，你要有心理准备，烈火即使出来，未必肯与你见面。”

荷生静静地说：“还有两年多时间，谁能预言未来。”

“说得很对，也许决定不再等待的会是你。”

“不，”荷生微笑，“那是你。”

言诺尴尬地看着她，“荷生，我永远说不过你。”

“嗳，你说得过表妹不就行了。”

这一天，樱花开了一树，不用风亦满枝乱颠，纷纷落下。

看门人正把落花扫到小径两边，看到荷生，微笑道：“春天到了。”

她点点头。

“孩子几时到?”

“下个月。”

“要格外留神。”

“谢谢你的关怀。”

她开启大门，看到一封信。

荷生并没有特别留神，她并没有即时捡起它，因为她此刻的身材，做蹲下的动作已经不十分方便了。

荷生先去打开窗户，放些新鲜空气进屋。

然后沏一杯热茶，慢慢喝着。

胎儿似乎有点不安，又似努力尝试在有限的空间内转动身躯。

荷生感到一阵剧痛，她失手掉了杯子，猛然记起医生的吩咐，连忙做深呼吸，松弛手足。

辛苦了五分钟，那种剧痛停顿下来，她拿起电话，与医生联络，医生说："你尽快到医院来吧，我随即赶来，春光明媚，恐怕小客人等不及要出来看看这个世界了。"

荷生一时不知道应该收拾些什么，看到杯子滚在地上，便用手托着腰，慢慢蹲下拾起它，它的旁边便是那封信，荷生亦顺带将之拣起放在桌上。

她取出卡片，打电话给烈家的司机。

"我是夏小姐，我想从公寓到医院去，你们可方便来接我？"

"十分钟即到。"

荷生道了谢。

她对刚才那剧痛犹有余悸，呆坐桌旁。

她低下头，看到白信封上写着她的名字，忽然之间，荷生察觉，这不是一封退信，也不是一封广告信，这是一封私人信件。

字迹完全陌生。

她轻轻拆开，信上短短三行字，她的名字之后，留了许多白，像是表示一个人的沉默，不知话该从何说起，然后，那人这样写：你信中的白字，也实在太多了一点。

荷生愕然，信，什么信？接着一个签名映入她的眼帘：烈火。

荷生发呆，不知是虚是实，是梦是真，随即想起，原来她从来没有见过烈火的签名，他们之间根本没有时间去发掘这些细节。

这会不会是什么人的恶作剧？

荷生不住抚摸着白信纸上的签名。

这时听见敲门声："夏小姐，夏小姐，有车子来接你。"

荷生抹一抹额角的汗珠，起身去开门，那封信紧紧握在手中。

门外是一脸笑容的言诺，"夏小姐，你准备好没有？"

荷生连忙拉住他，"言诺，言诺，你来看，这是谁的签名。"

言诺一看，"烈火！"

"这是烈火的笔迹？"

"的确是。"

荷生松下一口气来。

言诺明白了，他什么也不说，只是扶着荷生的手臂出门。

他感觉到有一股喜悦自荷生的手臂传过来，直达他的体内，连带感应了他，后来言诺觉得不对，荷生正紧皱眉头，歪曲着五官，正尽力忍痛，这股喜悦来自何处?

言诺忽然明白了，这快乐来自胎儿，是她，她在雀跃，她在鼓舞。

言诺轻轻对她说：“你有什么故事要告诉我?”

即使有，也不在上一代的篇幅之内了。

此刻，司机将车子飞驰到医院去，她的母亲手中，紧紧握着她父亲的一封来信。

一个希望。